Advanced Topics in Science and Technology in China

Zhejiang University is one of the leading universities in China. In Advanced Topics in Science and Technology in China, Zhejiang University Press and Springer jointly publish monographs by Chinese scholars and professors, as well as invited authors and editors from abroad who are outstanding experts and scholars in their fields. This series will be of interest to researchers, lecturers, and graduate students alike.

Advanced Topics in Science and Technology in China aims to present the latest and most cutting-edge theories, techniques, and methodologies in various research areas in China. It covers all disciplines in the fields of natural science and technology, including but not limited to, computer science, materials science, life sciences, engineering, environmental sciences, mathematics, and physics.

More information about this series at http://www.springer.com/series/7887

Hao Zhou · Kefa Cen

Combustion Optimization Based on Computational Intelligence

ZHEJIANG UNIVERSITY PRESS
浙江大学出版社

Springer

Hao Zhou
Zhejiang University
Hangzhou, Zhejiang
China

Kefa Cen
Institute for Thermal Power Engineering
Zhejiang University
Hangzhou, Zhejiang
China

ISSN 1995-6819 ISSN 1995-6827 (electronic)
Advanced Topics in Science and Technology in China
ISBN 978-981-10-7873-6 ISBN 978-981-10-7875-0 (eBook)
https://doi.org/10.1007/978-981-10-7875-0

Jointly published with Zhejiang University Press, Hangzhou, China

The print edition is not for sale in China Mainland. Customers from China Mainland please order the print book from: Zhejiang University Press

Library of Congress Control Number: 2017963634

Printed on acid-free paper

This Springer imprint is published by the registered company Springer Nature Singapore Pte Ltd.
part of Springer Nature
The registered company address is: 152 Beach Road, #21-01/04 Gateway East, Singapore 189721, Singapore

Preface

This book focuses on the combustion optimization. Coal still remains as the main energy resource in China, there are so many coal mines and their characteristics vary a lot, on the other hand, the load of the utility boilers changes frequently due to the increasing renewable energy usage. How to obtain the high combustion efficiency and the low pollutants emission simultaneously is an important issue. The gas combustion attracts more and more attentions due to the wide gas turbine and gas-fired boiler employment in China, it also faces the similar problems.

The influence of combustion parameters on NO_x emissions and carbon burnout briefly is discussed first. Second, due to the complexity of boiler systems, theoretical models are extremely difficult to build so far. In order to solve the problem, this book uses the artificial intelligence method which includes support vector regression (SVR) method, artificial neural network (ANN), and computational fluid dynamics (CFD) to obtain alternative models. Finally, this book raises the progress and equipment of online combustion optimization system and puts forward the future prospect of combustion optimization.

This book can be referred by researchers, engineers, and managers in the field of thermal engineering, and can also be used as an auxiliary material for teachers and students in colleges and universities.

I am greatly indebted to the clarifications and advice of several colleagues who took the time to proofread this book and suggest changes. Special thanks go to the colleagues and students in my research group, they are Kang ZHOU, Mingxi ZHOU, Yuan LI, Jiakai ZHANG, and Xutao GUO, thanks a lot for their contributions.

This work was supported by National Basic Research Program of China (2015CB251501), National Natural Science Foundation of China (51476137). Project No.51621005 supported by the Innovative Research Groups of the National Natural Science Foundation of China is also greatly appreciated.

"Combustion brings us heat, electricity, and long trip by plane, we can make combustion more efficient and clean."

Hangzhou, China

Hao Zhou
Kefa Cen

Contents

About the Authors

Prof. Hao Zhou received his Ph.D. from Zhejiang University in 2004. He is currently the Deputy Director of State Key Laboratory of Clean Energy Utilization at Zhejiang University and Director of the Zhejiang University—University of Leeds joint research center for sustainable energy. His research interests include combustion optimization, low pollutant combustion technology for utility boilers, and neural network and support vector machine modeling methods. He has published over 50 academic papers and filed seven patents in the areas of combustion pollutants control and combustion optimization since 2000.

Prof. Kefa Cen is a member of the Chinese Academy of Engineering. He received his Ph.D. from Moscow Industrial Technology University and is an expertise in clean coal combustion and gasification, poly-generation and comprehensive utilization of energy resources, as well as biomass gasification and bio-oil. He is currently the Director of the Institute for Thermal Power Engineering at Zhejiang University and Chairman of the Chinese Society of Power Engineering's International Cooperation & Exchange Committee. He is also Editor-in-Chief of the Journal of Zhejiang University (Engineering Science) and the Journal of Renewable Energy. He has published over 800 academic papers and 15 books.

List of Figures

List of Tables

Chapter 1
Introduction

Abstract In this chapter, the background and the key problems are presented, like the emission of nitrogen oxides (NO_x) and the level of unburned carbon. Selective Catalytic Reduction (SCR) and Selective Non-Catalytic Reduction (SNCR) are conducted to reduce the NO_x emissions. Artificial intelligence methods are used to solve the complexity of boiler system. The characteristic of coal combustion and the parameter of unburned carbon content are discussed in this chapter. Later, coal combustion optimization is proposed. The outline of the book is recommended at last.

Keywords Coal combustion optimization · Coal-fired power generation
Nitrogen oxides (NO_x) emission · Selective Satalytic Reduction (SCR)
Selective Non-catalytic Reduction (SNCR) · Support vector regression
(SVR) method · Artificial neural network (ANN) · Computational fluid dynamics
(CFD) · Global optimum

1.1 Background

Thermal power is China's dominant power generating capacity and accounts for three-quarters of the total capacity. Of the thermal generating capacity, more than 90% is coal-fired [1]. The emission of nitrogen oxides (NO_x) during coal combustion is a significant pollutant source to the environment. As environmental problems become more serious, the problem of NO_x emission is receiving increasing attention. On the other hand, the level of unburned carbon in fly ash is an important factor affecting the efficiency of pulverized coal-fired boilers, especially those equipped with low NO_x burners. High carbon content of fly ash will cause excessive heating of superheater and reheater tubes at the furnace outlet and may cause tube explosion. There are many old-designed coal-fired utility boilers in China. These boilers with low thermal efficiency generate rather high NO_x emissions. The NO_x emissions of these boilers may be reduced by installing the flue gas treatment equipment such as Selective Catalytic Reduction (SCR) and Selective

H. Zhou and K. Cen, *Combustion Optimization Based on Computational Intelligence*, Advanced Topics in Science and Technology in China,
https://doi.org/10.1007/978-981-10-7875-0_1

Non-catalytic Reduction (SNCR), however, the costs are high. Recently, combustion optimization has been proved to be an effective way to reduce the NO_x emissions and unburned carbon in fly ash in coal-fired utility boilers by carefully setting the operational parameters of boilers. Some achievements are made in their work, however NO_x emissions and carbon burnout are considered in isolation in their studies. It is well known that the air distribution implemented in low NO_x burners would induce high levels of unburned carbon in fly ash. Therefore, NO_x emissions and carbon burnout should be taken into account simultaneously. Due to the complexity of boiler systems, theoretical models are extremely difficult to build so far. However, alternative models could be obtained by using the artificial intelligence method [2–7]. Among them, the support vector regression (SVR) [5, 6] method, artificial neural network (ANN), [2, 4] and computational fluid dynamics (CFD), which have the advantages of global optimum, simple structure, and good generalization properties and have attracted wide spread attention.

1.2 Coal Combustion

1.2.1 General Process of Coal Combustion

The general process of pulverized coal combustion can be described as [8]:

(a) coal is heated, moisture in coal is separated;
(b) coal is heated continuously, volatile component separates out and causes inflammation first;
(c) coke is ignited, the remnants of volatile component separate out continuously, coke and volatile burn out together;
(d) ash forms.

1.2.2 The Duration of Coal Combustion

The time of volatile component separating out and burning completely takes over 10% barely of the whole combustion length, about 0.2–0.5 s. The time of coke burning out occupies more than 98% of whole process, about 1–2.8 s. The coke holds 60–95% of the heat released in total duration. Ignition process depends on the content of dry ash-free basis of coal, and the burnout process lies on the coke combustion speed mainly. According to practical experience, the length of burnout extends when the coal has a longer ignition time.

1.2.3 The Characteristic of Coal Combustion

The details of the coal combustion process are very complex, the main characteristics of several stages are discussed as follows.

The pulverized coal absorbs heat and its temperature rises, the heat source is derived from the gas in furnace, where the temperature reaches up to 1300–1600 °C. The high temperature gas heats the new fuel and raises its temperature by means of convection, radiation, and heat conduction. Moisture in coal separates out first, coal gets dried, and its temperature raises higher with the evaporation of moisture. In the temperature range of 120–450 °C, for different types of coal, volatile in coal separates and the rest solid state material forms coke.

Ignition temperature of combustible volatile gas is low, volatile gas causes inflammation when heated to 450–550 °C if oxygen is abundant, and at the same time, burning gas heats the coke. The coke absorbs the energy from the local high temperature area of burning volatile gas and high temperature flue gas in furnace, its temperature raises, and inflammation occurs when reaching ignition point, then an amount of heat is released.

After most of the coke has burnt out, the internal ash will impact combustion process. Internal ash of coke particle is distributed homogeneously in fuel, during the combustion process from the outer sphere to core gradually, the internal ash in outer sphere will envelop the core coke, forming a layer of ash shell, even dregs shell. This shell prevents oxygen scattering toward coke surface and prolongs the burnout process. Therefore, the influence of ash on combustion process is mainly on internal ash, and most of the external ash could not obstruct the burnable layer to be ignited individually.

Ignition temperature of pulverized coal airflow changes with coal fineness. The finer pulverized coal is, the faster pulverized coal would be heated. Therefore, the ignition time of pulverized coal can be shortened. This is because the finer coal possesses larger surface for combustion reaction.

It is difficult to discuss the coal combustion progress in furnace because the fineness, the description, and duration of volatile differ from each other. On some cases, when the thin pulverized coal particles have burnt out, the coarse ones just extract volatile gas.

Experiments have found that temperature rising rate of coal combustion in furnace is very speedy, reaching (0.5–1.0) × 104 °C/s, it can achieve 1500 °C in 0.1–0.2 s. In this condition, the combustion of volatile and coke cannot be distinguished remarkably, they proceed almost at the same time.

Experience has shown that pulverized coal with high amount dry ash-free basis is easy to be ignited and burned out. Volatile kindles rapidly than coke, and separation of volatile can accelerate the acreage for coke and oxygen reaction and increase the coke reactivity. Volatile components determine pulverized coal ignition, the greater the amount of volatile, the more heat is released in volatile combustion, which warms up coke sufficiently and aggrandizes contact with oxygen.

1.3 Carbon Burnout

With the developing demand for high efficiency of the utility boilers, more attention is paid to the unburned carbon content in the fly ash from the high capacity tangential firing boiler [9]. Fly ash is a final product of pulverized coal power plant after furnace, it depends on the coal type during combustion. But the unburned carbon content in the fly ash, one of the main contents of fly ash, is complicated and it is affected by many factors such as coal character, boiler's load, air distribution, boiler style, burner style, furnace temperature, excess air ratio, pulverized coal fineness, the uniformity of the air and coal distribution, etc.

The carbon content is normally in the range of 2–15 wt% carbon. Excessive amount of residual unburned carbon purport a significant energy loss and low combustion efficiency. A high carbon content also presents difficulties in disposal of ash. So in this way, the carbon burnout or unburned carbon predictor and measurer can assist the operator with carbon burnout adviser. But in fact, unburned carbon measurement cannot be calculated simply, it is usually implemented in entitative boiler to fumble the optimal operation condition to minimize unburned carbon.

The first one is a direct but expensive approach in which carries out combustion experiments in test furnace or operation plants. The second is based on computational fluid dynamics codes. The third involves predicting combustion behavior by using additives associated with the nature coals.

Published simple indices for burnout prediction include fuel ratio, mean vitrinite reflectance, heat value, and volatile matter of coals. A maceral index is developed to evaluate the burnout of coals and blends and is reported to be in good agreement with experimental data in pilot test facility [10].

Workload of entitative test is time consuming and short of operation conditions. Variety of factors would affect carbon burnout interlap and lead to difficulties in data analysis. Meanwhile design condition cannot be guaranteed on account of coal type and operational factors alteration, departure of combustion adjustment makes it difficult to obtain minimum unburned carbon. Relatively thermal losses can be obtained referring to related calculation formula.

CFD is an effective method to predict the performance and carbon burnout behavior of utility boilers. It has the potential to become an important design tool to help the engineers optimize the operating conditions, improve the heat rate, reduce pollution emission, improve the design of new boilers, and also evaluate the retrofit of old boilers. Generally, simulations of power plant boiler include several indispensable computational models. These comprehensive models form the basis of the numerical solution of the multidimensional differential equations for conservation of mass, energy, and momentum. However, building CFD models tends to be comprehensive, time-consuming project, requiring significant computing facilities, which are mainly used for new plant design or old plant retrofit. It is difficult for the operators to estimate the combustion performance under a changing operating condition, or optimize the carbon burnout by using CFD simulation.

How to acquire the relation between operation conditions and unburned carbon shows great significance to combustion optimization.

1.4 Coal Combustion Optimization

Coal combustion optimization is complex, but also shows a great significance to power plant. Coal combustion optimization emphasizes diversely for different objects: low pollution emission, high boiler efficiency, preventing slagging, etc. These objects interact with each other, subjected to economy and safety. Most of optimization carried out stress on specific goal, ignoring the global best state. Coupled parameters influencing coal combustion are nearly innumerable, mathematical formula of parameter influence cannot be obtained at present, these fluctuating nonlinear parameters are also formidably to be measured during combustion. Incalculable efforts have been invested into coal combustion optimization area to solve different problems, offline mathematical tools for model have been utilized for optimization, fuzzy mathematics as well as the neural network have been applied in online optimization. The popular methods combine artificial model tools with CFD based on chemical mechanism to simulate and optimize.

Coal combustion optimization is carried out mainly from fuel and air. Specifically, coal combustion can be controlled by experimental measurement of coal type, air distribution and combustion angle in furnace. And conclusion based on experiments provides theoretical basis for modeling. Taking NO_x emission optimization for example. Studies simulate how the secondary air temperature influences NO emission in 350 and 800 °C through a simplified chemical model, comparison between results and measurement shows NO emission is reduced in high temperature air aura. This finds a new though for NO_x emission optimization, but the optimal result is rough, object is too simple and this method is dependent on wind power measurement intensely. Expert system diagnoses combustion state by fuzzy logic and the corresponding models, its core is knowledge base and inference engine, containing expert-level experience and knowledge essentially. But expert system is subjected by dealing complex problem because of knowledge representation. Artificial algorithm (neural network, support vector machine, etc.) combined with optimization algorithm (GA, PSO, ACO, etc.) has been proved to be an effective way to optimize coal combustion [3, 4, 7].

1.5 Outline of the Book

The book is organized as follows: Chap. 2 will discuss the influence of combustion parameters on NO_x emissions and carbon burnout briefly; Detailed modeling methods for combustion characteristics including experimental method, CFD and computational intelligence method will be introduced in Chap. 3; In Chaps. 4 and 5, neural network and support vector machine methods modeling of combustion characteristics will be analyzed; Examples of optimization combining modeling methods and optimization algorithms will be introduced in Chap. 6; Progress and equipment of online combustion optimization system will be raised in Chap. 7; In Chap. 8, future prospect of combustion optimization will be put forward.

References

1. Lam PM, Shiu A. Efficiency and productivity of China's thermal power generation. Rev Ind Organ. 2004;24:73–93.
2. Zhou H, Cen KF, Fan JR. Modeling and optimization of the NO_x emission characteristics of a tangentially fired boiler with artificial neural networks. Energy. 2004;29:167–83.
3. Zheng LG, Zhou H, Cen KF, Wang CL. A comparative study of optimization algorithms for low NO_x combustion modification at a coal-fired utility boiler. Expert Syst Appl. 2009;36 (2):2780–93.
4. Zhou H, Cen KF, Mao JB. Combining neural network and genetic algorithms to optimize low NO_x pulverized coal combustion. Fuel. 2001;80(15):21633–2169.
5. Wu F, Zhou H, Tao R, Zheng LG, Cen KF. Combining support vector regression and cellular genetic algorithm for multi-objective optimization of coal-fired utility boilers. Fuel. 2009;88:1864–70.
6. Bloch G, Lauer F, Colin G, Chamaillard Y. Support vector regression from simulation data and few experimental samples. Inf Sci. 2008;178(20):3813–27.
7. Li R, Cui YM, He H, Wang HN. Application of support vector machine combined with K-nearest neighbors in solar flare and solar proton events forecasting. Adv Space Res. 2008;42(9):1469–74.
8. Fan Y. Guoluyuanli. Beijing: China electric power press; 2008.
9. Styszko-Grochowiak K, Gołaś J, Jankowski H, Koziński S. Characterization of the coal fly ash for the purpose of improvement of industrial on-line measurement of unburned carbon content. Fuel. 2004;83(13):1847–53.
10. Helle S, Gordon A, Alfaro G, et al. Coal blend combustion: link between unburnt carbon in fly ashes and maceral composition. Fuel Process Technol. 2003;80(3):209–23.

Chapter 2
The Influence of Combustion Parameters on NO_x Emissions and Carbon Burnout

Abstract In this chapter, the influence of combustion parameters on NO_x emissions and carbon burnout is discussed briefly. On the one hand, the effects of coal type, chemical equivalent and residence time, temperature, moisture and ash content, air dynamic field and flame species, particle size, boiler load, and OFA nozzle on NO_x emissions are investigated. On the other hand, the influence of combustion parameters, such as the operational conditions of the boiler and the coal properties, OFA nozzle, primary air, inner secondary air and outer secondary air and swirling intensity on carbon burnout are studied.

Keywords Control of NO_x emissions · Efficient utilization of coal
Unburned carbon in ash · Influencing factors of NO_x formation and decomposition
Influence of combustion parameters on carbon burnout · Operational conditions of the boiler · OFA nozzle

2.1 Introduction

The emissions of nitrogen oxides during coal combustion are the major environmental problems. The control of NO_x emissions from coal-fired boilers is an international concern to the power generation industry. It is well known that NO_x emissions are dependent upon the combustion process, depending on ambient conditions, coal quality, and the condition of plant equipment. The emission of NO_x from boilers can be reduced significantly by modifying the combustion process.

Coal remains the primary energy resource in the world. One of the main objectives in coal combustion research is the development of technologies to help power plant designers and operators to perform clean and efficient utilization of coal. High levels of unburned carbon resulting from inefficient coal combustion lead to major environmental and economic consequences. Reductions in the levels of unburned carbon in ash are helpful to improve operational efficiency, and minimize solid waste for pulverized coal-fired boilers, especially those equipped with low NO_x burners. Due to the reduced mixing intensity and the formation of

H. Zhou and K. Cen, *Combustion Optimization Based on Computational Intelligence*, Advanced Topics in Science and Technology in China,
https://doi.org/10.1007/978-981-10-7875-0_2

fuel-rich zones under low NO_x combustion, the residence time of coal particles in an oxygen-rich environment decreases, resulting in the increase of unburned carbon content in the fly ash.

Unburned carbon in ash may be dominated by the fuel ratio, particle size, maceral composition, reactive maceral content, rank and combustion conditions. At present, there are three major ways to predict the burnout behavior of coals and blends, including experimental, computational and indices [1]. The first one is a direct but expensive approach, carrying out combustion experiments in furnaces or plants. The second is based on computational fluid dynamics (CFD) codes. The third involves predicting combustion behavior by using indices associated with the nature of coals. Published simple indices for burnout prediction include fuel ratio [2], mean vitrinite reflectance [3], and heat value, volatile matter of the coals [4]. A maceral index is developed to evaluate the burnout of coals and blends in Ref. [5] and is reported to be in good agreement with the experimental data in a pilot test facility.

2.2 Influence of Combustion Parameters on NO_x Emissions

Because of the complexity of the mechanism of NO_x formation and decomposition, combustion parameters' influences on these process are many and complicating. In summary, there are some main factors:

(1) coal type (2) chemical equivalent and residence time (3) temperature (4) moisture and ash content (5) air dynamic field and flame species (6) particle size (7) boiler load (8) OFA nozzle.

(1) Coal type

Under the condition of no grading combustion, the relationship between NO_x emission and coal volatile and nitrogen content is very clear. The higher the coal volatile and nitrogen content, the higher the NO_x emission. However, this relationship tends to be not clear under staged combustion conditions.

HCN and NH_3 formed are subsequently oxidized to NO or N_2, depending on the prevailing conditions. Coal properties have an effect on NO reduction by the partitioning of nitrogen between the char and the volatiles under staged combustion, since the volatile nitrogen is mainly sensitive to the sub-stoichiometric conditions in the flame. Under unstaged conditions, the conversion of volatile-N to NO is more than the conversion to NO of char-N. Higher volatile content or lower rank therefore gives higher NO emissions at unstaged conditions. During staged combustion, NO emission is reduced, and the amount of reduction increases with volatile content. In this case, a higher volatile content gives lower NO emissions. In general, a difference in rank causes also a difference in temperature and residence time in the fuel-rich zone, influencing the nitrogen release and reaction rates indirectly.

A relation between the nitrogen content of a coal and the NO produced can be found, although it is not more than a very rough trend: more nitrogen gives more NO (Fig. 2.1). The nitrogen content does therefore not give much of an indication of the NO concentrations to be expected. Morgan found that under unstaged conditions a clear correlation existed between the volatile-N and NO emission. At staged conditions, however, the relation was not clear and the variation in emissions was much smaller between different coals. The tests were conducted in a 2.5 MW combustor, with external and internal air-staging concepts. The optimum burner configuration was shown to be a function of coal type.

(2) Chemical equivalent and residence time

The best chemical equivalent which leads low NO_x emission is when the excess air coefficient is about 0.7. High oxygen concentration results in the increasing NO_x emission. The NO_x comes from the oxidation of HCN. The heterogeneous reduction of NO_x is also an important factor. Although it is a slow response, rich combustion areas tend to have very small char particle diameter. The char can reduce NO_x but the effect is limited. CO has a promoting effect on the heterogeneous reduction of NO_x. For the anthracite, when the residence time is 1 s in lower level of combustion, NO_x begins to fall until the excess air coefficient reaches 0.45. The test shows that the lower the level of combustion, the longer the residence time, and the more it is beneficial to reduce NO_x emissions.

(3) Temperature

The major nitrogen forms of gas products released at 600 °C were HCN and NH_3, whereas only a slight amount of N_2 was formed. When the temperature was increased up to 1000 °C, N_2 was the dominant product regardless of coal type. There was no relationship between N_2 formation and nitrogen content in coal or the amount of volatile matter released at 1000 °C.

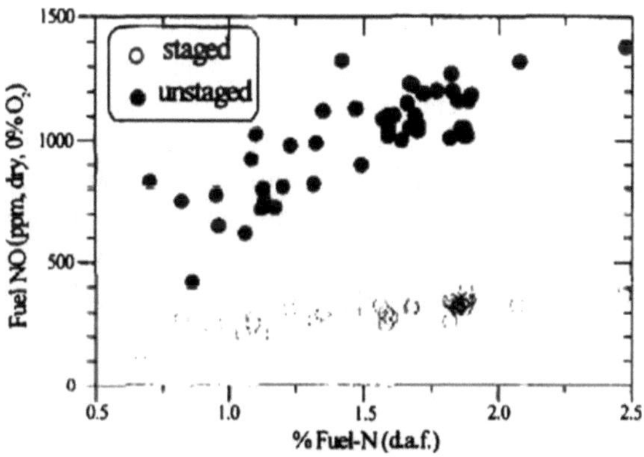

Fig. 2.1 Fuel-NO produced during staged and unstaged combustion of a suite of coals as a function of the fuel-N content [1]

At 700–1100 °C, the production of HCN increases dramatically with temperature. At higher temperature (1100–1400 °C), N_2 generates extensively and sometimes even more than 50%.

(4) Moisture and ash content

Moisture has an effect on NO_x emissions. Due to the reduction effect of H_2O, moisture leads to the reduction of NO_x emission. In normal and staged combustion conditions, with the increase of ash content, NO_x concentration and unburned carbon have increased at furnace exit. With the ash increasing, the effect of NO_x reduction which is brought by the staged combustion becomes weak, but the fly ash carbon content increased significantly.

(5) Air dynamic field and flame species

Aerodynamics has great influence on the formation of NO_x. When coal pyrolysis in the fuel-rich conditions, the flow field gradually mixed with the air in the combustion zone is conducive to achieve low NO_x emissions. Recirculation zone for NO_x emissions is also very important. The recirculation zone can enhance the coal combustion. Near the exit of the burner, the coal flames ignited quickly. It is very important to reduce NO_x emissions. Secondary air needs to be mixed after the devolatilization. Grading air supply reduces the combustion temperature. But it also brings combustion instability and increases the fly ash carbon content. The pulverized coal concentrate of primary air is also conducive to the establishment of low NO_x emission flame structure.

(6) Particle size

Particle size has a complicated influence on NO_x emission. Some experiments show that large particles lead to low NO_x emissions. Small particles lead to high NO_x emissions (an increase of about 30%). Some experiments show that the particle size has no effect on NO_x emissions. Other studies show that an intermediate particle size leads to the highest NO_x emission. When increasing or decreasing the size, NO_x emission can be reduced. Deviations of these tests may be related with the test apparatus. When all the combustion air is fed from the burner, small particles lead to higher NO_x emissions. Because small particles cause fast jet diffusion and are mixed with secondary air quickly, it leads to high NO_x emission. For the result about large particles with high NO_x emission, it may be caused by the large particles with a short residence time in a fuel-rich zone. Primary air are used several times than secondary air in the experiments where the particle size has no effect on NO_x emission. Therefore, all primary air is through the recirculation zone. The particle size has no effect.

(7) Boiler load

In contrast to the results obtained with nitrogen retention in the char, some differences have been observed concerning the evolution of hydrogen cyanide and ammonia. HCN formation has been found to be nearly unaffected by elevated pressure, whereas the NH_3 release has strongly decreased. The greatest changes have occurred at low heating rates because of increased residence time of the primary devolatilization products (Fig. 2.2). According to the overall reaction, ammonia formation should be promoted under high pressure.

$$2HCN + 6H_2 \rightleftharpoons 2NH_3 + 2CH_4$$

However, it has been found that this reaction only has effect below 873 K. From this aspect, the conversion of ammonia mainly to N_2 must dominate with increasing pressure. Apart from the catalytic decomposition of ammonia, reactions with hydrogen radicals have to be considered to understand the chemical mechanisms.

The NO_x emission decreases with the decreasing of the boiler load, as illustrated in Fig. 2.3. The NO_x emission decreases much more rapidly in the range of

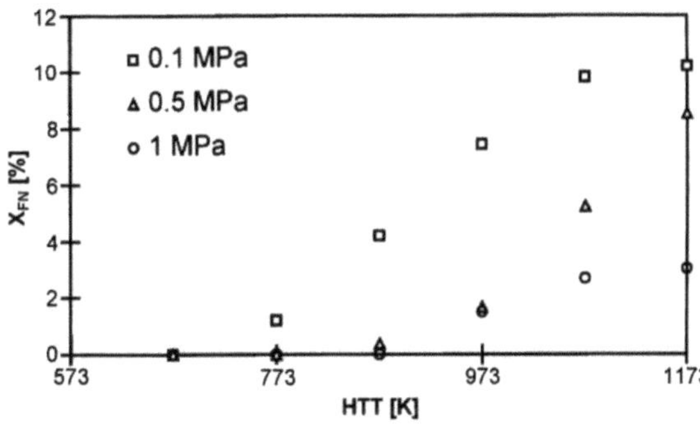

Fig. 2.2 NH_3 evolution at elevated pressure (CG lig., 0.25 K/min) [3]

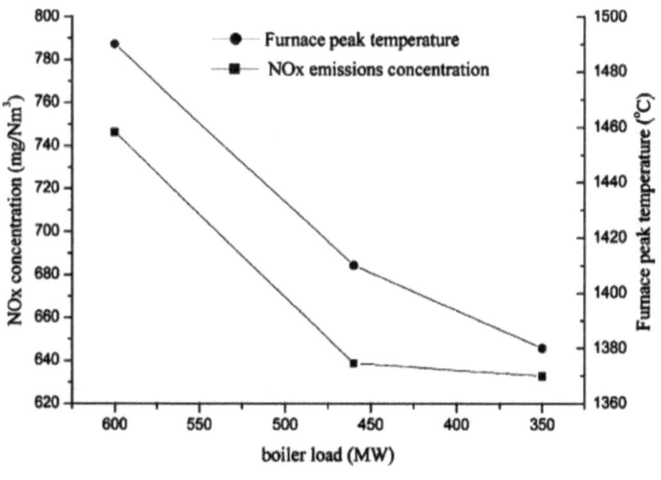

Fig. 2.3 The influence of boiler load on the combustion performance [4]

600−480 MW than that in the range of 480−350 MW. It implies that the thermal NO_x occupies a rather large proportion of the total NO_x release under the higher boiler load. The furnace peak temperature measured under the 600 MW boiler load was 1490 °C, while that of 480 and 350 MW were 1410 and 1380 °C, respectively.

(8) OFA nozzle

Many existing studies show that 60–80% of the fuel-NO_x emission of coal combustion is generated from volatile-N. So the process control of volatile pyrolytic reaction and volatile combustion is effective to reduce NO_x emission. With fractional air combustion method, the NO_x emission can be significantly reduced because the coal in the main combustion zone burned out with less oxygen. However, combustion with less oxygen may increase the carbon content in the fly ash. OFA is distributed above the main combustion zone to make the residual char fully burn out. So it is essential to investigate the effects of the relative position of OFA on NO_x emission and carbon content in fly ash. The relative position of OFA nozzle is the quotient of the distance between the OFA nozzle and the main burner with the length of the furnace.

Figure 2.4 shows the effects of the relative position of OFA on NO_x emission. There is a high agreement among the effects of OFA nozzle on NO_x emission of three different experimental coals. NO_x emission values are up to the maximum when the relative position of OFA nozzle is 0.32, and the NO_x emission value is relatively smaller when the relative position of OFA is 0.22 and 0.42. When the OFA nozzle is located close to the main burner, the OFA can reduce the temperature and the combustion speed of the fuel flow, which control the NO_x generation and emission. When the OFA nozzle is located far from the main burner, the same

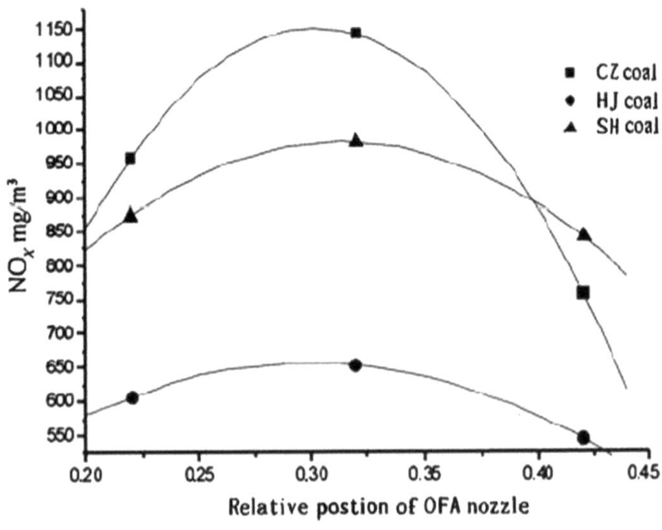

Fig. 2.4 NO_x emission value varied with the relative position of OFA nozzle

effect occurs because insufficient oxygen lowers the NO_x generation. When the relative position of OFA nozzle is 0.32, OFA is supplied at the time that the volatile pyrolytic reaction is just completed and the coke begins to burn. Thus, an optimum local condition is formed for NO_x generation and the NO_x emission values is at the maximum.

2.3 Influence of Combustion Parameters on Carbon Burnout

(1) The operational conditions of the boiler and the coal properties

Tests 2.1 and 2.2 were conducted to test the effect of the secondary air distribution pattern on the carbon burnout (Tables 2.1 and 2.2).

During the experiments, coal fineness was kept constant, and main gas species concentrations were continuously monitored in the boiler outlet prior to the air heater. Solids samples were withdrawn from the flue gas by an isokinetic cyclone dust collector. The carbon burnout data reported here are average values over several hours of stable operation. Pulverized coal samples were extracted from the burner supply lines with isokinetic multi-point sampling system. The measurements were performed a week later after the boiler switched to the pure test coal, to make the boiler conditions constant. Only pure test coal was fired during a test. The measured unburned carbon contents in fly ash are summarized in Table 2.3.

(2) OFA nozzle

Combustion with less oxygen may increase the carbon content in the fly ash. OFA is distributed above the main combustion zone to make the residual carbon content to fully burn out. The relative position of OFA nozzle is the quotient of the distance between the OFA nozzle and the main burner with the length of the furnace.

The effects of the relative position of OFA nozzle on the burn-off rate are shown in Fig. 2.5. When the relative position of OFA nozzle is 0.33, the carbon content in the fly ash is at the minimum. The function of OFA is to supply oxygen for combustion of the residual carbon content. If it is located close to the main burner, the OFA will lower the temperature and the combustion speed of the fuel flow, and increase the carbon content in fly ash. If the OFA nozzle is located far from the main burner, the same effect occurs because the combustion speed is slow.

(3) Primary air, inner secondary air, and outer secondary air

The effects of primary air ratio on carbon content in fly ash are shown in Fig. 2.6. Since the pulverized coal distribution of this burner is lean outside and rich inside along the diameter of burner nozzle, the thick pulverized coal is burned at the high-temperature backflow zone, and with the increase of ratio of primary air, the oxygen is more sufficient to accelerate the speed of combustion. This is effective to

Table 2.1 The boiler operating condition

Case	Total fuel rate (t/h)	Total air rate (t/h)	The opening value of the coal feeder (%)					The opening value of the secondary air damper (%)						The opening value of OFA (%)
			A	B	C	D	E	AA	AB	BC	CD	DE	EF	
1	238.6	2367	67.77	68.80	70.60	66.17	67.97	65	65	65	65	65	65	60
2	237.1	2350	68.93	68.83	69.93	66.07	66.17	65	65	65	65	65	65	60
3	231.7	2328	64.37	69.23	68.93	67.80	61.34	65	65	65	65	65	65	60
4	232.7	2293	64.73	69.63	69.00	68.10	61.37	65	65	65	65	65	65	31
5	228.7	2283	63.13	68.17	68.23	67.23	60.70	65	65	65	65	65	65	0
6	240.4	2337	67.27	72.97	72.80	67.10	60.27	65	40	40	80	80	85	60
7	242.6	2343	70.30	73.77	73.10	67.50	61.10	65	65	65	65	65	85	60
S	241.0	2344	68.00	72.77	72.77	67.20	60.37	65	65	65	65	65	65	60
9	240.6	2367	70.97	73.37	73.07	67.30	59.57	65	65	65	65	65	65	60
10	180.7	1875	65.50	66.57	66.60	60.50	0	45	45	45	45	25	45	60
11	143.0	1520	49.77	52.73	53.30	47.27	0	45	45	45	45	45	45	48
12	239.2	2321	69.07	73.57	73.27	67.60	60.43	65	65	65	65	65	65	60

Table 2.2 The boiler operating condition

Case	OFB	O_2 in flue gas (%)	The coal quality (on "as received" basis)						The air flow rate through the mill (*t*/h)					The nozzles tilt[a]
			C_{ar} (%)	H_{ar} (%)	O_{ar} (%)	N_{ar} (%)	$Q_{net,ar,p}$ (kJ/kg)	V_{ar} (%)	A	B	C	D	E	
1	40	3.307	64.51	3.99	6.53	1.56	26,069	28.11	111.4	105.8	106.6	110.2	102.2	0.5
2	100	3.325	64.51	3.99	6.53	1.56	26,069	28.11	111.6	106.1	104.0	109.2	103.2	0.5
3	0	3.022	64.51	3.99	6.53	1.56	26,069	28.11	109.0	107.3	105.1	105.4	103.9	0.5
4	0	3.042	64.51	3.99	6.53	1.56	26,069	28.11	98.54	106.4	99.65	105.8	106.1	0.5
5	0	2.762	64.51	3.99	6.53	1.56	26,069	28.11	109.7	108.1	104.6	107.6	105.2	0.5
6	40	3.100	57.89	3.17	9.94	0.96	23,677	24.75	104.3	118.9	117.5	106.7	106.0	0.5
7	40	3.079	57.89	3.17	9.94	0.96	23,677	24.75	104.1	110.4	114.0	106.7	105.5	0.5
8	40	3.091	57.89	3.17	9.94	0.96	23,677	24.75	102.5	110.8	115.9	107.7	105.6	0.6
9	40	3.232	57.89	3.17	9.94	0.96	23,677	24.75	101.9	112.2	117.9	107.6	104.2	0.4
10	0	3.922	64.51	3.99	6.53	1.56	26,069	28.11	108.8	112.4	103.5	114.6	20.45	0.5
11	0	4.732	64.51	3.99	6.53	1.56	26,069	28.11	112.4	110.4	101.4	118.1	17.30	0.5
12	40	3.089	57.89	3.17	9.94	0.96	23,677	24.75	102.8	110.2	116.5	106.7	105.8	0.5

[a]The nozzles are tilted horizontally to 50%. In case 8, the nozzles are tilted upwards to 60%. In case 9, the nozzles are tilted downwards to 40%

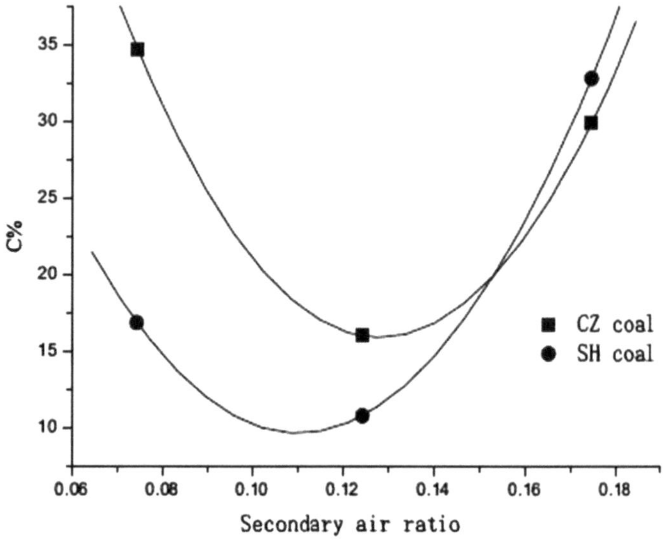

Fig. 2.7 Carbon content in fly ash varied with inner secondary air ratio [5]

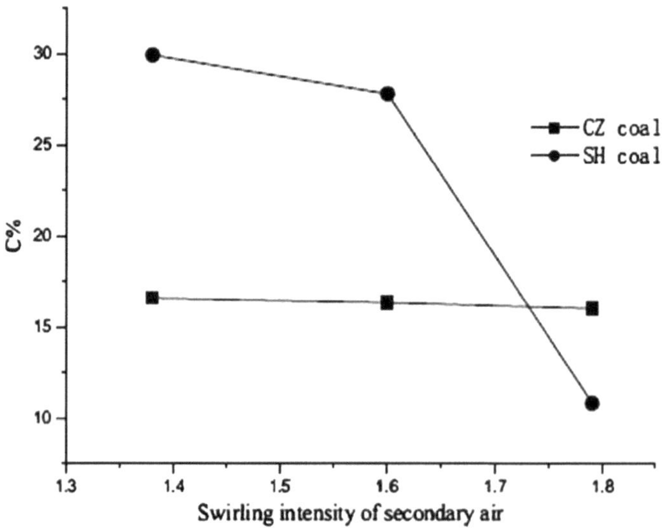

Fig. 2.8 Carbon content in fly ash varied with swirling intensity

than 0.125, with the increase in the ratio, the carbon content cannot be fully burned out, so the carbon content in fly ash will increase.

The effects of outer secondary air ratio on carbon content in fly ash are shown in Fig. 2.8. The results show that the outer secondary air is helpful for reducing the carbon content in fly ash.

(4) Swirling intensity

Figure 2.8 shows the effects of swirling intensity on carbon content in fly ash. Carbon content in fly ash decreases with the increase in inner and outer secondary air because increase of swirling intensity enhances backflow, turbulence, and mixture of the fuel flow, which is helpful for the volatile pyrolytic reaction and volatile combustion.

References

1. Van der Lans RP, Glarborg P, Dam-Johansen K. Influence of process parameters on nitrogen oxide formation in pulverized coal burners. Prog Energy Combust Sci. 1997;23(4):349–77.
2. Wu Z, Ohtsuka Y. Remarkable formation of N_2 from a Chinese lignite during coal pyrolysis. Energy Fuels. 1996;10(6):1280–1.
3. Friebel J, Köpsel RFW. The fate of nitrogen during pyrolysis of German low rank coals—a parameter study. Fuel. 1999;78(8):923–32.
4. Zhou H, Cen K, Fan J. Modeling and optimization of the NO_x emission characteristics of a tangentially fired boiler with artificial neural networks. Energy. 2004;29(1):167–83.
5. Xue S, Hui S, Liu T, et al. Experimental investigation on NO_x emission and carbon burnout from a radially biased pulverized coal whirl burner. Fuel Process Technol. 2009;90(9):1142–7.

Chapter 3
Modeling Methods for Combustion Characteristics

Abstract In this chapter, through the study of the complex combustion process, different influence factors are exposed. Experimental method and CFD methods are compared, respectively, with economic and accuracy. The former focuses on the detailed and accurate information about coal combustion, including ash content, fusion temperature, flame temperature, and flue gas. Many CFD methods, such as turbulence model and radiative heat transfer model, are introduced to understand their appropriate operating condition. Due to the convenience to analyze large and complex data, CFD methods are widely applied in combustion simulation. Computational intelligence method based on combustion studying is also proposed in this section.

Keywords Combustion optimization methods · Coal combustion characteristics
Ash composition analysis · A novel flame image processing technique
Reconstruct 3-D flame temperature · Comprehensive numerical simulation in the furnace · Turbulence models · Gas-phase turbulent combustion
Non-premixed combustion model · Computational intelligence method

3.1 Introduction

Optimization methods are usually divided into three orientation: (1) simplified model based on experience and experiment, with which to control fuel supply and air distribution; (2) optimization according to the models calculated by fuzzy algorithm, genetic algorithm, and other artificial algorithm; (3) precisely optimization based on CFD and chemical mechanism. The first simple way takes few parameters into account; it adhibited in early optimization and is fused with the second new method. The second optimization method can deal with numerous nonlinear inputs, and vigoroso calculation and searching ability ensure online optimization onset. It has been widely applied in many prospects. The third method demonstrates the accurate combustion process in furnace; its development potential is limited to the inconformity in simplified mechanism and practical reaction; and the result is not very reliable and time-consuming, and its application is under reaching.

H. Zhou and K. Cen, *Combustion Optimization Based on Computational Intelligence*, Advanced Topics in Science and Technology in China,
https://doi.org/10.1007/978-981-10-7875-0_3

3.2 Experimental Method

3.2.1 Experimental Methods of Coal Combustion Characteristics Study

To varying degrees, coal-fired units in China have an adaptability problem of combustion device and fuel characteristics. The common issues include difficult ignition, unstable combustion, low combustion efficiency, fouling and slagging, over temperature tube-burst and frequent maintenance, etc. Coal is a kind of organic mineral polymer. The difference in coal-forming age and geological conditions will lead to a different physical and chemical characteristic of coal. Therefore, coal shows a different combustion performance. The distributions of coal in China are widespread and distribution unbalance. Fuel coal is various from lignite to anthracite in coal properties. To realize the utilization of the limited coal resources better, it is necessary to study the coal combustion characteristics.

3.2.1.1 Coal Combustion Characteristics

Pulverized coal combustion performance has an important influence on boiler design, alteration, and combustion adjusting. Coal combustion characteristics include many aspects, such as ignition characteristics, volatile release characteristics, burnout features, pyrolysis properties, surface and pore structure characteristics, fouling and slagging properties, wear properties and pollutant emissions, etc. Coal combustion characteristics can be evaluated based on the volatile yield to a certain degree. However, in some cases, it may not match with the actual operation. Especially, burning the poor coal has similar volatile yield, and the opposite result may be obtained. This is due to the fact that coal combustion is not only affected by the chemical factors but also by the physical factors. And different coal structures may have the same chemical composition [1]. Considering the safety and economic operation of the boiler, the most important is the stability of the coal fire, combustion, and slagging. Practices have also proved that if the three issues are handled well with a reasonable designed boiler, the boiler would be operated safely and economically. Coal combustion characteristic indexes are shown in Table 3.1 [2].

3.2.1.2 Experimental Methods

The data used in coal combustion indexes above can be obtained by experiments as follows: industrial analysis, elemental analysis, ash melting point, ash composition analysis, thermal gravimetric analysis, etc. [3].

Table 3.1 Coal combustion characteristic index

Indexes		Definitions	Comments
Ignition stability	Devolatilization index D	$D = (\mathrm{d}w/\mathrm{d}t)_{\max}/T_{l\max} \times \Delta T_{1/3}$	$(\mathrm{d}w/\mathrm{d}t)_{\max}$: the maximum release rate of volatile $T_{l\max}$: temperature of maximum heat release rate $\Delta T_{1/3}$: temperature regions of $(\mathrm{d}w/\mathrm{d}t)/(\mathrm{d}w/\mathrm{d}t)_{\max} = 1/3$
	Ignition property index F_z	$F_Z = (V_{\mathrm{ad}} + M_{\mathrm{ad}})^2 C_{\mathrm{ad}} \times 100$	Judged by industrial analysis of coal The larger the value, the better the fire performance
	Ignition stability index W_I	$W_I = a/T_i + b/T_{l\max} + c(\mathrm{d}w/\mathrm{d}t)_{\max}$	T_i: ignition temperature a, b, c: calculation constants The larger the value, the better the fire performance
Burnout characteristic	Combustion characteristic index S	$S = (\mathrm{d}w/\mathrm{d}t)'_{\max}(\mathrm{d}w/\mathrm{d}t)'_{\mathrm{mean}}/(T_i^2 \times T_h)$	$(\mathrm{d}w/\mathrm{d}t)'_{\max}$: maximum combustion rate of coal tar $(\mathrm{d}w/\mathrm{d}t)'_{\mathrm{mean}}$: mean combustion rate of coal tar T_i: ignition temperature, T_h: burnout temperature The bigger the S value, the better burnout, the higher economy effect
	Burnout characteristic index W_B	$W_B = 10/(aG + bT_{2\max} + c\tau + \mathrm{d}\tau')$	G: the amount of post-burn combustion $T_{2\max}$: temperature of maximum coal tar combustion rate τ, τ': coal and char burnout time a, b, c, d: calculation constants The bigger the F_B value, the better coal burnout

(continued)

Table 3.1 (continued)

Indexes		Definitions	Comments
Ash fusibility	Ash melting characteristic	DT, ST, FT $T = (DT + 4ST)/5$	DT: deformation temperature; ST: softening temperature; FT: flowing temperature Ash fusion method is intuitive, the higher melting point, the more difficult slagging
	Ash component	Alkali acid ratio: $B/A = \frac{Fe_2O_3 + CaO + MgO + Na_2O + K_2O}{SiO_2 + Al_2O_3 + TiO_2}$ Silica alumina ratio: $S/A = SiO_2/Al_2O_3$ Silicon ratio: $G = \frac{100 \times SiO_2}{SiO_2 + Fe_2O_3 + CaO + MgO}$ Iron calcium ratio: $F/C = Fe_2O_3/CaO$ Dolomite ratio: $D = \frac{CaO + MgO}{CaO + MgO + Fe_2O_3 + Na_2O + K_2O}$ $B/A \times S$	Ash component method is indirect and reliable. With the increase of B/A and S/A, ash melting point is decreased drastically. In contrast, the melting point is increased with the increase of G
	Ash viscosity	Viscosity–temperature curve T_{CV} $T250$ Plastic viscosity temperature range	10–1000 Pa s T_{CV}: Critical viscosity temperature T_{250}: temperature at a viscosity of 250 Pa s $T_{CV} - T_{1000}$
	Comprehensive slagging index	$W_s = xC_{t2} + y/(C_{B/A} - C_G) + zC_{S/A}$	$C_{t2} = 12 - 0.007t_2, C_G = 22.5 - 0.29G$ $C_{B/A} = 0.45 = 5.2B/A, C_{S/A} = 1.2S/A - 0.9$ x, y, z: calculation constants The larger the F_s value, the stronger coal slagging
Deposition	Deposition index	$H_w = B/A \times Na_2O$	The larger the H_w value, the more serious soot formation
Wear	Wear index	$H_M = A(S_iO_2 + 0.8Fe_2O_3 + 1.35Al_2O_3)$	A: ash (as received basis) The larger H_M value, the more serious wear

Ash fusibility

The ash melting temperature is an index for estimating ash bonding in grate furnace. It has also been used for slagging tendencies discrimination in solid-state slag pulverized coal furnace. It is still one of the indispensable basic data for the design and operation of a large-scale boiler.

Ash melting properties influence the furnace operating conditions. Pyramid method is generally used to test the ash fusibility. The method has some advantages such as convenience, intuition, high efficiency, etc. But the disadvantage is the inevitable subjective error. Make the ashes into a cone which has an equilateral triangle bottom; the length of the cone bottom is 7 mm, and the cone height is 20 mm. The ash cone is heated at a specified heating rate. And we get three ash fusion temperatures that represent the ash fusibility according to the state change of ash cone, as shown in Fig. 3.1:

(1) Deformation Temperature, DT. The cone top becomes round or begins to tilt.
(2) Softening Temperature, ST. The cone top bends to cone bottom or it shrinks into a spherical.
(3) Flowing Temperature, FT. The cone turns into a liquid and can flow on the plane.

The coal ash occurs with a slight caking property under a temperature higher than DT. The formation of slag is slowly. When the temperature is lower than DT, a loose dry ash deposition will be formed on the heating surface. Once the temperature reaches ash softening temperature, there will be a lot of slag. Under the FT, ash will flow along the wall surface or drip directly.

Operating experience shows that boiler with dry bottom furnace is less likely to slagging under the case ST > 1350 °C. Furnace outlet temperature should be lower than ST to avoid furnace exit slagging, and the difference between them set aside at least 50–100 °C [4].

Ash composition analysis

The coal ash content has a certain influence on the formation of contamination in the furnace. Ash composition is very complicated, mainly including CaO, MgO, Fe_2O_3, Na_2O, K_2O, SiO_2, Al_2O_3, and TiO_2. Constants, semi-micro, capacity, and atomic absorption methods are commonly used experimental methods. The

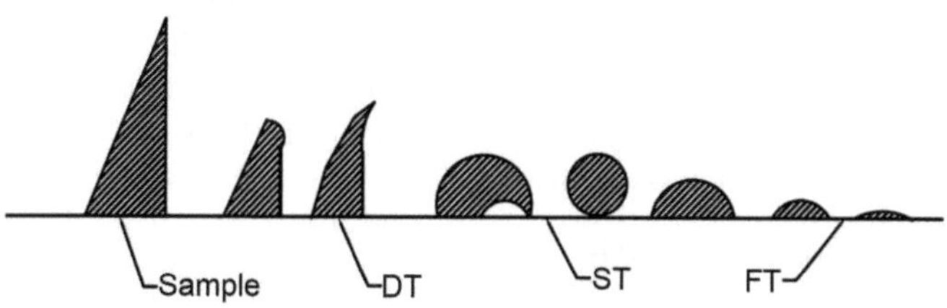

Fig. 3.1 The state change of ash cone

appropriate method is chosen based on the actual situation. GB/T 1574-2007 gives detailed information about reagent, apparatus, analytical procedures, results calculation, and the precision of methods [5].

Elemental analysis and industrial analysis

Coal is a mixture of organic and inorganic components. Its molecular structure is very complicated. Both elemental analysis and industrial analysis are usually used to determine the percentage content in coal. Organic compounds in coal will gradually decompose during heating, so it is impossible to measure coal organic compounds directly based on existing analysis method. Therefore, the elemental composition of coal is used to represent the characteristics of coal organic compounds.

The main chemical elements in coal include carbon (C), hydrogen (H), oxygen (O), nitrogen (N), and sulfur (S). The total of five materials with ash and moisture content together is amount to 100. Coal elemental composition is tested by burning. First, sample the raw coal and send sample to the laboratory. Then, test the sample using the following procedures: sample weighing, adding chemical reagent, titration, and making use of atomic absorption spectrophotometer. It is noteworthy that the methods described above are strictly in accordance to GB. In the field of combustion laboratory, off-line coal elemental analysis instrument and methods mainly include coal element analyzer, inductively coupled plasma atomic emission spectrometry (ICP-AES), and X-ray fluorescence spectrometry (XRF).

Elemental analysis is quite complex in operation. Generally, only industrial analysis is made in power plant. Industrial analysis components include moisture, ash, volatile matter, and fixed carbon. The total amount of these four components is 100. Industrial analysis composition is not the fuel original composition. It is obtained by drying, heating, or burning the coal under some certain conditions. Decomposition and transformation of extremely complex composition in coal occur during these processes.

Thermal analysis

Thermal analysis technology began in the 1887 Le Chatelier research on clay, and it has been a history of more than 100 years. Robert Auster made use of thermogravimetric analysis only in studying steel and other metals in 1899. Thermogravimetric analyzer was studied out by Hohda in 1915. In the 1958, Freemean Carroll studied the measurement methods of thermogravimetric (TG). In the 70s, thermal analyzer tended to be perfect in terms of automation and microanalysis. Currently, thermal analysis is one of the most commonly used methods in coal combustion characteristics at home and abroad [6–10], and the schematic diagram of thermal analysis system is shown in Fig. 3.2. Thermogravimetry, differential thermal analysis, and differential scanning method are the most widely applied in the thermal analysis technologies.

(a) **Thermogravimetric analysis**

Thermogravimetric analysis is a technique that measures the relationship between mass and temperature at a programed heating rate by thermobalance. The main

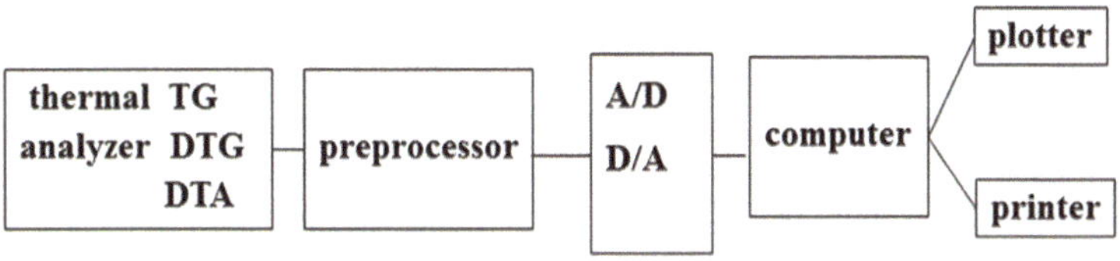

Fig. 3.2 The schematic diagram of thermal analysis system

components of thermobalance include record balance, stove, process control temperature system, recorder, and support device. With the increase of temperature, the sample weight loss rate d*w*/d*t* changes constantly until it is burnt out. The curve recording changes of d*W*/d*t* with temperature are called "combustion distribution curve." Combustion distribution curve of U.S. B&W company is shown in Fig. 3.3. The experiment conditions are as follows: 300 mg coal sample was heated at 15 °C/min heating rate in the air.

The generally used controlling temperature methods have the following three ways: cooling or heating samples at a programed heating rate; keeping the sample at a constant temperature; and controlling the temperature at a some sort of program cycle. Then, thermogravimetric analysis curves (TGA) and derivative thermogravimetric curves (DTG) are obtained. The coal combustion characteristics are studied through an analysis of its TG and DTG, such as ignition temperature,

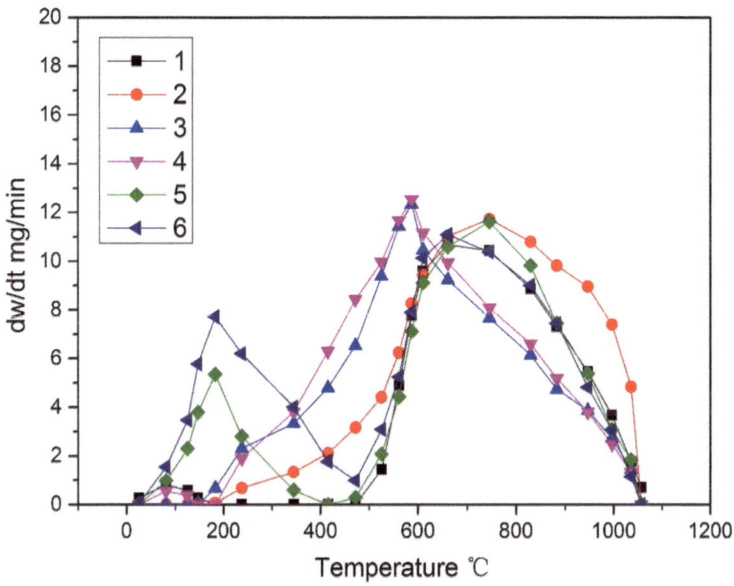

Fig. 3.3 Combustion distribution curve of B&W [10]. ① anthracite coal; ② lean coal; ③ low volatile bituminous coal; ④ low volatile bituminous coal; ⑤ inferior bituminous coal; and ⑥ lignite coal

burnout temperature, burnout time, maximum weight loss temperature, maximum weight loss rate, etc. Different coals have different TG and DTG curves, and also have different characteristic values. Therefore, different coals have different ignition and combustion performances. The combustion characteristics can be well evaluated through these indexes formed by combining the data described above. The typical DTG curve of thermal analysis is shown in Fig. 3.4, and the meanings of feature points are as follows: A represents water begin to evaporate; B is the maximum weight loss rate of water; C represents volatile starts to precipitate; D is the maximum weight loss rate of volatiles and the corresponding temperature; E is a specific temperature (700 °C), and the weight loss of combustion before this temperature is ΔW700; F is the maximum weight loss rate of fixed carbon; and G presents the coal is burnt out.

In recent years, the combination of thermal balance, gas chromatograph (GC), and Fourier transform infrared spectroscopy makes it possible to study the changes of components during coal combustion [11, 12].

It is worth noting that some thermal analyzer due to limits of heat and mass transfer, coal sample is difficult to burn in the analysis of low volatile coal and anthracite. Thermal analyzer is characterized by intuitive and poor repeatability in the analysis of the kinetic parameters of coal and coke. In particular, kinetic parameters obtained by different thermal analyzer may be orders of magnitude difference [13]. Therefore, quantitative evaluation of coal combustion characteristics by thermal analyzer remains to be further work.

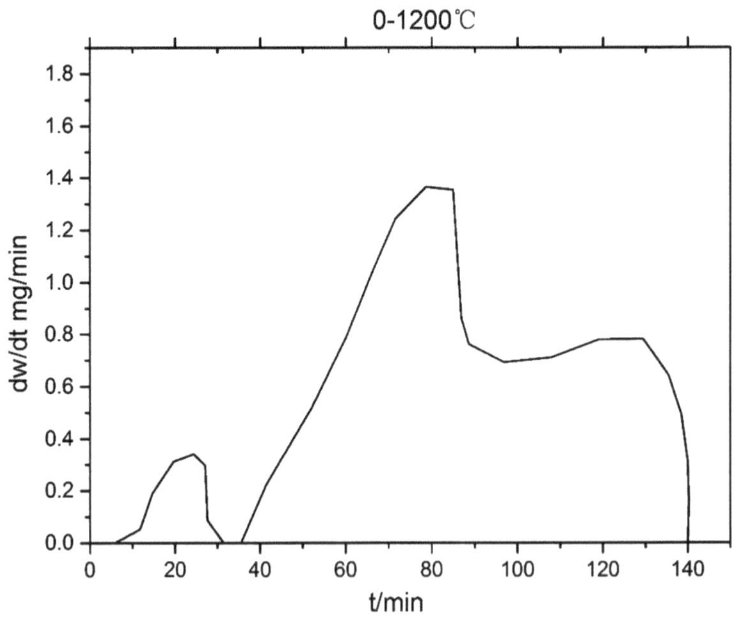

Fig. 3.4 DTG curve of Datong coal [10]

(b) **Differential thermal analysis (DTA)**

Differential thermal analysis is a technique that measures the relationship of temperature difference or d*T*/d*t* between sample and reference material (a kind of material on which no thermal effect occurs during a range of temperatures) at a programed heating rate. The temperature difference between sample and reference is recorded as a function of temperature and time during the experiment. Coal sample experiences physical and chemical changes associated with endothermic and exothermic phenomenon when heated or cooled. Differential thermal analysis is a method based on this nature of coal.

The basic principle of DTA is to put the coal sample and a reference material under the same conditions like heating or cooling. The coal temperature would be different from reference in this process, and then exothermic or endothermic peak could be obtained from the DTA curve.

The temperature difference between coal sample and reference is recorded as a function of temperature or time during the experiment. The mathematical expression is as follows:

$$\Delta T = T_s - T_r = f(T\ or\ \mathrm{t}),$$

where T_s, T_r represent the sample and the reference material temperature, respectively, T is the program temperature, and t is time.

In the DTA experiment, the sample temperature change is caused by the phase transition and the effect of reaction endothermic or exothermic effects, such as crystallization, melting, evaporation, decomposition reaction, and redox reaction. The reaction processes during coal combustion will be known well on the basis of DTA, especially after corresponding it with the TGA. Various processes include the water volatile combustion, fixed carbon burning, and coal burnout temperature.

(c) **Differential scanning (DSC)**

Differential scanning (DSC) is a technique that measures the relationship between input power difference and temperature. The DSC is divided into two kinds based on the different measurement methods: power compensated differential scanning and thermal differential scanning. The temperature difference between coal sample and reference is zero during analysis process in power compensated differential scanning. It is achieved by power compensation. The temperature difference between coal sample and reference is proportional to heat flow difference between these.

3.2.1.3 Test System of Coal Combustion

Drop tube furnaces (DTFS)

Drop tube furnaces (DTFS) which also called entrained flow reactor is one of the most basic and commonly used combustion test systems [14]. The British BRUC laboratory and the Australia CSIRO laboratory first conducted a large number of experiments to study coal combustion characteristics through DTFS in the 1960–1970s.

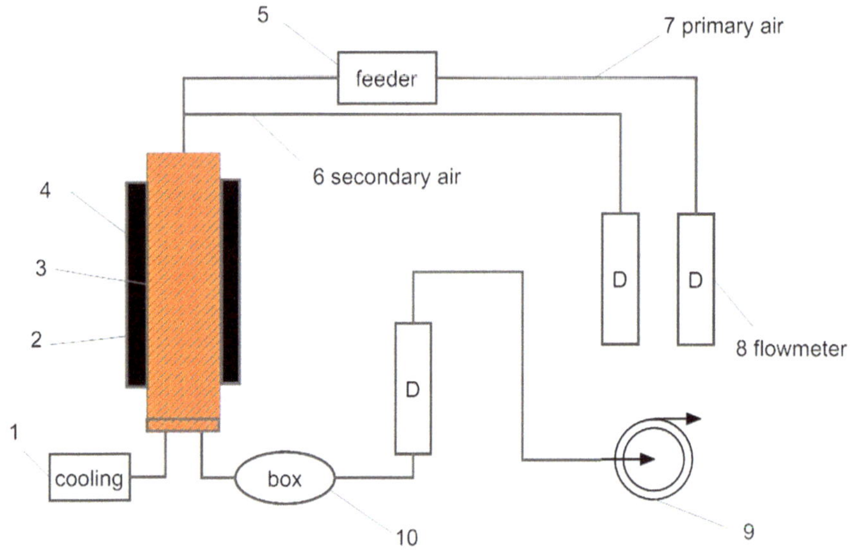

Fig. 3.5 The system of DTFS [14]. 1—cooling water inlet, 2—sampling tube, 3—reaction tube, 4—heating pipe and protective layer, 5—feeder, 6—secondary air, 7—primary air, 8—flowmeter, 9—pump, and 10—sampling box

DTF technology has been gradually developed since then [15–19]. DTFS system consists of feeder, reaction tube, sampling tube, and temperature controlling system, as shown in Fig. 3.5. The DTFS is usually used to study coal combustion process, coal burnout characteristics, SO_2 emissions, NO_x emission, and trace elements concentration. The main parameters of DTFS are listed in Table 3.2.
DTF characteristics are as follows:

(1) High heating rate: DTF is an important heating rate laboratory reactor [20, 21]. DTF heating rate is about 104–105 °C/s [22, 23] which has the same order of magnitude with the actual heating rate (105–106 °C/s) in combustion equipment and industrial gasification furnace. And both heating methods also use similar heating mode. The mainly heating modes of these two methods are convection heating. Therefore, the actual furnace thermal properties can be well simulated by DTF.
(2) Exclude the interaction between particles: Feeding rate of DEF feeder is very small which is about 0.5–5 g/min. The average size of coal particle is a few microns typically. Thus, pulverized coal carried by air enter into the furnace can achieve better separation. The agglomeration of particles is effectively prevented, which can largely eliminate the effect of particles and internal diffusion on the reaction.

Table 3.2 Parameters of some foreign DTFS

Pressure/ 10^5 Pa	Max. temperature/°C	Residence time/s	Feeding rate/ g min^{-1}	Reaction tube D × L/ mm × mm
1	1000	0.03–0.11	0.25–0.5	50 × 950
1	977	0.025–0.27		50 × 600
1	1427			63 × 2000
1	1827	0.005–0.20	0.12–0.18	51 × 460
1	1650	0.01–0.7	2.0–3.0	51 × 750
1	1700	0.2–4.0	0.01–8.3	
6–70	1000	0.06–1.7	0.2–0.5	40 × 550
1–16	1427	0.03–1.0	0.5–1.5	51 × 1200
1–80	1500			70 × 2000
1–25	1600		0.2–0.5	50 × 1000
1–25	1800	2.0–8.0		30/50 × 2050
1–16	1400	0.02–10.0	0.3–1.5	75 × 1500

(3) Dynamic performance: Solid-phase continuous flow technology used in DTF makes pulverized coal that has good dynamic performance in the furnace. The coal dynamic performance is closer to the actual condition of industrial device. Samples obtained from different axial positions along the tube make the study of coal reaction characteristics under different residence times possible.

(4) Suppress secondary reactions: The precipitates of coal particle will react again on the heated coal particles in cumulate coal. The secondary reaction affects the activity of remaining reactants. The number of secondary reactions is effectively reduced in DTF due to the better separation of coal particles. Meanwhile, the high heating rate makes the touching time between decomposition produces and heated coal shorter; this also helps to reduce the possibility of secondary reaction.

(5) Product analysis: Coal in the feeding device is enough to guarantee the continuous of experiment. Thus, it can provide a sufficient amount of reaction products to be analyzed.

One-dimensional furnace

According to the flow state, one-dimensional furnace can be divided into the following three kinds: plug flow reactor, laminar flow reactor, and entrained flow reactor. It is generally accepted that laminar flow reactor and load flow reactor is more suitable to simulate flame than DTFS. Some basic researches about coal combustion characteristics under different conditions can be made on one-dimensional furnace. Then, the influences of various parameters on combustion can be well understood in order to achieve combustion optimization. This combustion system can also be used to study formation mechanism of SO_2 and NO_x, therefore control the pollutant emissions fundamentally. At the same time, it can also provides experimental reference and verification for different combustion burner. Most of institutions engaging in coal combustion study have built similar one-dimensional furnace test beds such as the University of Utah and Brigham Young University in the U.S., Zhejiang University, Tsinghua University, and Xi'an Jiao Tong University in China [24–26]. A great deal of valuable research results about combustion chemical kinetics, NO_x emission, and particulate emissions are obtained.

One-dimensional furnace in the University of Utah is a 100 kW oxygen-enriched combustion facility, as shown in Fig. 3.6, and consists of vertical and horizontal sections. The vertical section includes heating region and non-heating region. Three groups of resistance wire heat the heating region mainly for the fire area. The non-heating region is the radiation heat transfer area. Continuous sampling can be carried out nine times along the vertical tube. A quartz glass is used as the observation window to facilitate visualization. There are also four sampling points along the horizontal section. Meanwhile, horizontal section is also equipped with eight groups of heat exchanger to low exhaust temperature.

Notably, analysis of coal combustion characteristics must be performed in standard conditions. Otherwise, different analytical results obtained are not comparable. Nowadays, the most widely used is thermal analysis test based on coal. DTF and one-dimensional furnace are generally used to compare with industrial applications.

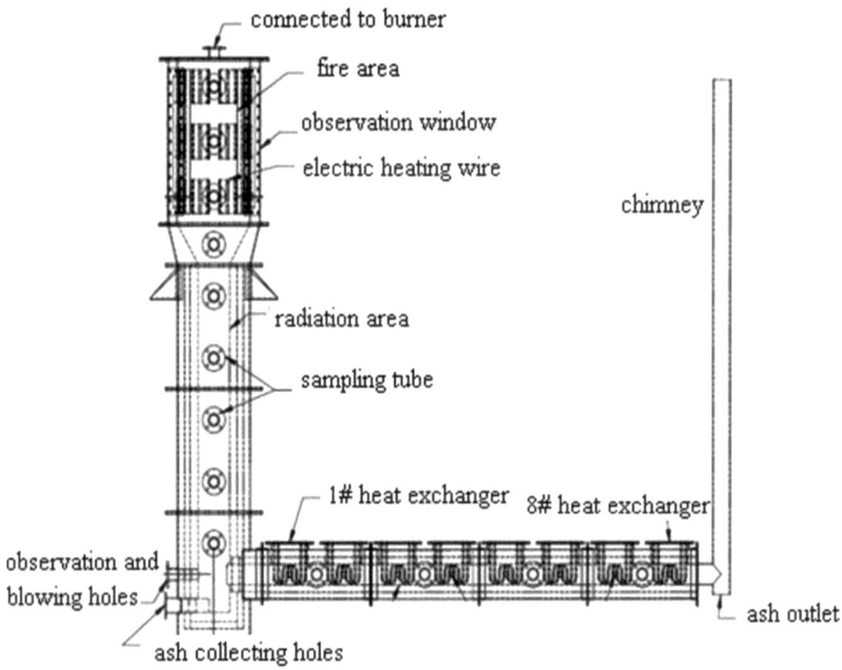

Fig. 3.6 One-dimensional furnace [25]

3.2.2 Flame Temperature Measurement

The online real-time measurement of flame temperature is important for combustion optimization and pollutants control in boiler combustion adjustment test. At present, there are many temperature measurement methods. The methods are divided into two kinds contact method and noncontact method based on the relationship between the sensor and tested material. Contact temperature measurement is a method that two contacted objects realize thermal equilibrium after a long enough time. If choose one as a standard thermometer , then the other temperature can be obtained. Noncontact temperature measurement is a method that the standard reference used as a thermometer and tested object do not touch. The temperature is obtained through detecting thermal radiation or some other feature quantities associated with temperature.

The typical contact temperature measurement method is thermocouple. A thermocouple consists of two dissimilar conductors in contact, which produces a *voltage* when temperature of the two junction parts is different. The temperature is obtained by the calibration of the heated voltage. However, the sensor usually can only measure a point in flame. The flame temperature distribution is unknown. The online real-time measurement is difficult to achieve because of the large thermal inertia and the hard establishment of thermal equilibrium. The higher flame temperature and complicated operation conditions easily reduce the thermocouple lifetime and measurement accuracy.

Noncontact temperature measurement consists of measuring the thermodynamic parameters and thermal radiation character [27]. The method will not destroy the flame temperature field, and measuring response time is short. Therefore, noncontact method is easy to realize online real-time dynamic measurement. However, an optical window is needed to spread thermal radiation spectrum. The passing rate of thermal radiation may be uneven due to local contamination of window which increases the difficulty of flame temperature measurement.

With the development of computer and semiconductor technology, the video signal of camera can be converted into digital image. The digital image can be quantitative analyzed by computer, which makes the subsequent analysis and automatic monitoring possible. Currently, more and more researchers combine the digital image processing technology with thermal radiation principles to measure flame temperature distribution and other parameters. Charge couple device (CCD) is a new radiation temperature measurement method that uses modern photoelectric detection technology, digital image processing technology, and thermal radiation theory. Besides the advantages of radiation thermometry, it also has the following advantages:

(1) The measured object emissivity and medium selective absorption have little effect on measurement result. Measuring accuracy is high.
(2) Thermal inertia is small and response time is short. Online real-time dynamic measurement can be achieved.

(3) The measurement system is suitable for corrosive object because it is not affected by measured object.
(4) Environmental temperature has little effect on the device, so it can be used in harsh condition of strong electromagnetic interference.
(5) It can be used to measure temperature of mobile and rotating object.
(6) It can provide surface temperature distribution of measured object. A pseudo-color display of temperature can be achieved.
(7) Measurement data are long-distance transmitted through the network to achieve remote monitoring and resource sharing.

3.2.3 Flue Gas Analysis

Sampling

The flue gas composition measurement is of importance in power plant boiler. It can assess the following parameters: the degree of fuel combustion, combustion condition, combustion characteristics of different location, the relationship between operation condition and pollutant emission, etc.
Flue gas sampling is the key to ensure the accuracy of composition analysis. The general requirements of flue gas sampling in furnace are as follows: continuous automatic sampling, the sampling position avoid from a chemical reaction as far as possible, measuring point should be far away from the local resistance, boundary layer can not be sampled. In order to measure the concentration of components in flue gas correctly, sampling must be carried out under the constant velocity condition. Constant velocity, namely that the suction velocity of sampling probe, is equal to the around gas velocity.Planck constant

Flue gas composition analysis technology [28, 29]

(1) Gas Chromatography

Chromatography is a *laboratory technique* for the *separation of mixtures*. The mixture is dissolved in a fluid called the mobile phase, which carries it through a structure holding another material called the stationary phase. The tube containing stationary phase is chromatographic column. The chromatography is divided into gas chromatography (GC) and liquid chromatography based on the state of mobile phase. GC is generally used in the research field of coal combustion. And GC has been widely used in power plant because of its simple structure, easy operation, high sensitivity, and short analysis time.

GC analysis is performed according two steps: (1) medium containing the sample to be analyzed flows through the stationary phase to separate components in sample, and then the separated components are brought into detection room; (2) the single component is measured successively in the detection room. The output of a set of signal according to time is chromatogram.

(2) Mass spectrum analysis

A mass spectrum is a histogram representing the distribution of ions by mass in a sample.

A mass spectrometer consists of four components [2]: a sample injecting system, an ion source, a mass analyzer, and a detector. Gas sample is injected into mass spectrometer by sample injecting system. Then, the ionizer converts a portion of the sample into ions. An extraction system removes ions from the sample, which are then trajected through the mass analyzer and onto the detector. The differences in masses of the fragments allow the mass analyzer to sort the ions by their mass-to-charge ratio. The detector measures the value of an indicator quantity and thus provides data for calculating the abundances of each ion present.

(3) Infrared analysis

Most of the coal combustion products can be tested by infrared analyzer. As known in physics, energy transition will be occurred on a molecule when it is radiated by a certain wavelengths. The energy difference and the radiation frequency are in accordance with the following relationship:

$$E_2 - E_1 = h\gamma,$$

where $E_2 - E_1$ is the energy difference, h is Planck constant, $h = 6.625 \times 10^{-27}$ erg/s, and γ is the light frequency. The external radiation will be absorbed by the molecule when the radiation frequency is in accordance with the above formula. And a selective energy level transition occurs on the molecule. Different molecules absorb different light of various frequencies, which is named selective absorption. The absorbed radiation frequency is characteristic frequency or wavelength.

Infrared spectroscopy is mainly used to study the compound with dipole moment change in vibration. Almost all organic molecules and gas molecules are absorbed in the infrared range except some single atoms such as He, Ne, Ar, and some homonuclear diatomic molecules such as N_2, O_2, H_2, etc. The infrared absorption band wavelength position and intensity represent the features of molecule structure. So it can be used to identify molecule structure and make qualitative analysis. The absorption intensity of an absorption band is related to the concentration of molecule components. Therefore, it can also be used to make quantitative analysis and purity identification [2].

(4) SO_2 analysis

Sulfur contained in coal is oxidized into SO_2 in the process of coal combustion. Monitoring the SO_2 emission is a powerful method to monitor the desulfurization efficiency of desulfurizer and SO_2 concentration. The classical approach of testing SO_2 is iodometry method. First, SO_2 is absorbed by mixture of H_3NSO_3 and $(NH_4)_2SO_4$. And then, the mixture is titrated by iodine standard solution. At present, there are also many other methods such as constant potential electrolysis, infrared absorption, UV fluorescence method, the solution conductivity method, and neutralization method. The principles and interference factors of some methods are listed in Table 3.3 [2].

(5) NO_x analysis

NO, NO_2, and some other nitrogen oxide are often referred as NO_x. The NO_x emission analysis in flue gas is a trace analysis of ppm level. And many of the measurement technologies have been developed such as nondispersive infrared absorption method (NDIR), ultraviolet absorption method, chemiluminescence method, etc. Since the characteristic absorption peaks of NO and H_2O are both in the vicinity of 5.3 μm, which has a great impact on the test results in the NDIR. So the NDIR technique is rarely used. Ultraviolet absorption method utilizes the characteristic absorption peak of gas components in the UV visible region to test the NO_x. The measurement range of chemiluminescence is wide. It is suitable for flue gas analysis. The NO_x concentration is measured through detecting a narrow region energy in the total emitted energy.

(6) Oxygen analysis

Oxygen measurement is very important in the coal-fired power plant. Oxygen in the flue gas is the result of excess air in the combustion process. Oxygen analyzer mainly includes thermomagnetic oxygen analyzer, zirconia oxygen analyzer, and constant potential electrolysis.

Experiments prove that oxygen is paramagnetic material, and its magnetic susceptibility is much larger than other gases. Therefore, the oxygen content can be tested through measuring the magnetic properties of the mixed gas. The generally used paramagnetic analyses are thermaomagnetic analysis, differential analysis, and automatic zero balance analysis.

ZrO_2 ceramic is a kind of solid electrolyte and has ionic conductivity. Current exists between the solution and gas containing different concentrations of chemical under the condition of the electrolytic. The current depends on the chemical potential difference. It is characterized by a simple structure and preprocessing system, a high sensitivity and resolution, a wide measurement range, and a short response time. Therefore, ZrO_2 analyzer has been widely used in most of the combustion control boiler in power plant.

Constant potential electrolysis is a portable analyzer developed in recent years. Typical battery used in this technology is a metal air battery consists of anode, electrolyte, and air cathode. The reaction occurs between positive and negative as follows:

$$2Pb + O_2 \rightarrow 2PbO$$

A current proportional to the oxygen content in the samples is produced in the above reaction. The sensors are usually not used for continuous operation.

Table 3.3 The principles and interference factors of some methods [2]

Method	Principle	Testing concentration range mg/m^3	Interference factor
Iodometry	$SO_2 + H_2O \rightarrow H_2SO_3$ $H_2SO_3 + I_2 + H_2O \rightarrow H_2SO_4 + 2HI$	100–5700	Positive interference: H_2S, NO_2 Negative interference: O_2
H_2O_2-Ba $(ClO_4)_2{\cdot}3H_2O$-$C1_6H_{11}AsN_2Na_2O_{10}S_2$	SO_2 is oxidized into H_2SO_4 by H_2O_2, then it is titrated in pink through $Ba(ClO_4)_2$, and the thorin is indicator used to judge the solution color change	1–80,000	Free ammonia, water-soluble cation and oxide
Neutralization method	$SO_2 + H_2O_2 \rightarrow H_2SO_4$ $H_2SO_4 + 2NaOH \rightarrow Na_2SO_4 + 2H_2O$	More than 1400	Other acidic materials
Constant potential electrolysis	$SO_2 + 2H_2O_2 \rightarrow SO_4^{2-} + 4H^+ + 2e^-$(standard oxidation potential $+0.17\,V$) $i = \frac{nFADC}{\delta}$, n–shifted electrons, F–Faraday constant, A–diffusion area, D–diffusion coefficient, δ–thickness of diffusion layer	0–5700 0–11,400 0–14,300	Positive interference: H_2S, Negative interference: NO_2
Solution conductivity method	$SO_2 + H_2O_2 \rightarrow H_2SO_4$		Positive interference: HCl, Cl_2, CO_2, NO_2, H_2S Negative interference: NH_3

3.2.4 Application Examples

Li [30] conducts an experimental investigation on the estimation of radiative properties and temperature distributions in a 670 t/h coal-fired boiler furnace by a portable imaging processing system. The portable system has been calibrated by a blackbody furnace. Comparing the equivalent blackbody temperatures measured by the portable system and the infrared pyrometer, the relative difference is less than 4%. The reconstructed pseudo-instantaneous 2-D temperature distributions in two cross sections can disclose the combustion status inside the furnace. The measured radiative properties of particles in the furnace can provide useful information for the calculation of radiative heat transfer and numerical simulation of combustion in coal-fired boiler furnaces. The preliminary experimental results show that this technology will be helpful for the combustion diagnosis in coal-fired boiler furnaces (Figs. 3.7, 3.8, and 3.9).

Zhou [31] introduces a novel flame image processing technique, and the 3-D temperature distributions of combustion in the pulverized coal-fired boiler furnace of a 200 MW power generation unit were visualized experimentally. With the assumption of gray radiation, multiple color flame image detectors were used to capture approximately monochromatic radiation intensity images under the visible wavelengths of red (R), green (G), and blue (B). A blackbody furnace was used to calibrate the color images. A way to calculate the radiation intensity using the

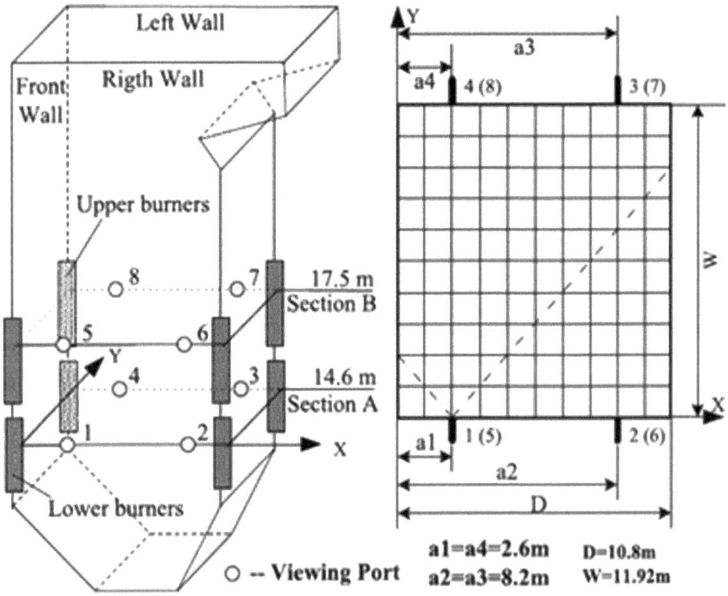

Fig. 3.7 Schematic of the furnace and the viewing ports, through which flame images were captured by the portable system by turns [30]

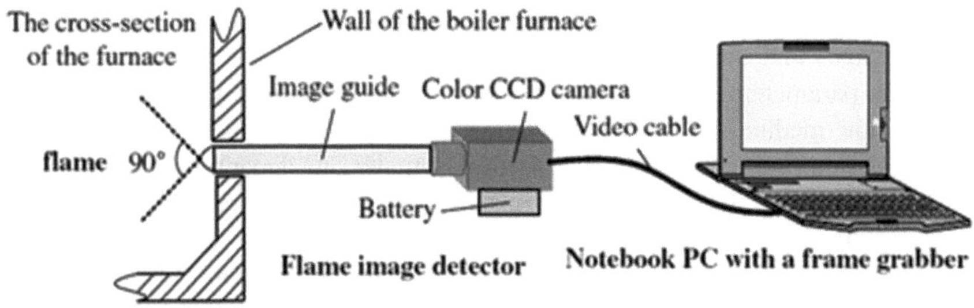

Fig. 3.8 Schematic of the portable image processing system [30]

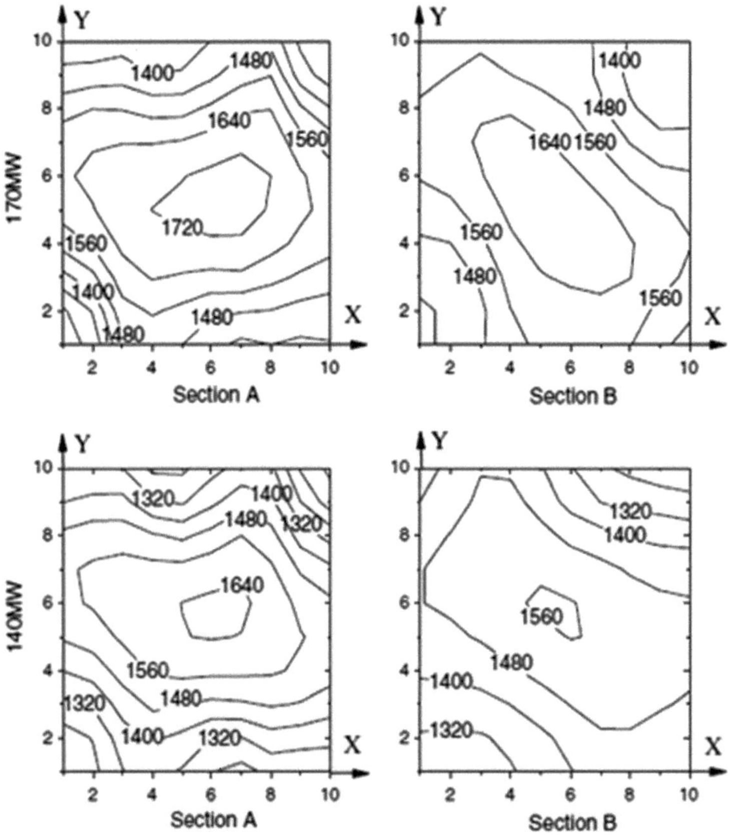

Fig. 3.9 Pseudo-instantaneous 2-D temperature distributions in sections A and B of the 670 t/h boiler furnace in two furnace loads of 170 MW and 140 MW, respectively [30]

Monte Carlo method was used to establish the relationship between the radiation flame images and the 3-D temperature distribution inside the furnace, given the radiative parameters such as the absorption and scattering coefficients of the combustion medium. A modified Tikhonov regularization method was used to reconstruct the 3-D temperature distribution from the 2-D flame temperature images transformed from the color flame images. Eight flame image detectors were mounted close to the four corners in four layers in different heights in the furnace; the 3-D temperature distribution along the height of the furnace was reconstructed from the eight flame images. The errors between some local temperatures visualized by the present system with those measured by a suction thermocouple were within ±5%, showing applicability of the method. The preliminary results show that this technology will be useful for the combustion diagnosis and control in coal-fired power generation plants (Figs. 3.10, 3.11, 3.12, and 3.13).

Gong [32] proposes a new combination of the optical sectioning tomography (OST) and two-color method to reconstruct the three-dimensional (3-D) temperature distribution of the gasifier by employing a single charge-coupled device

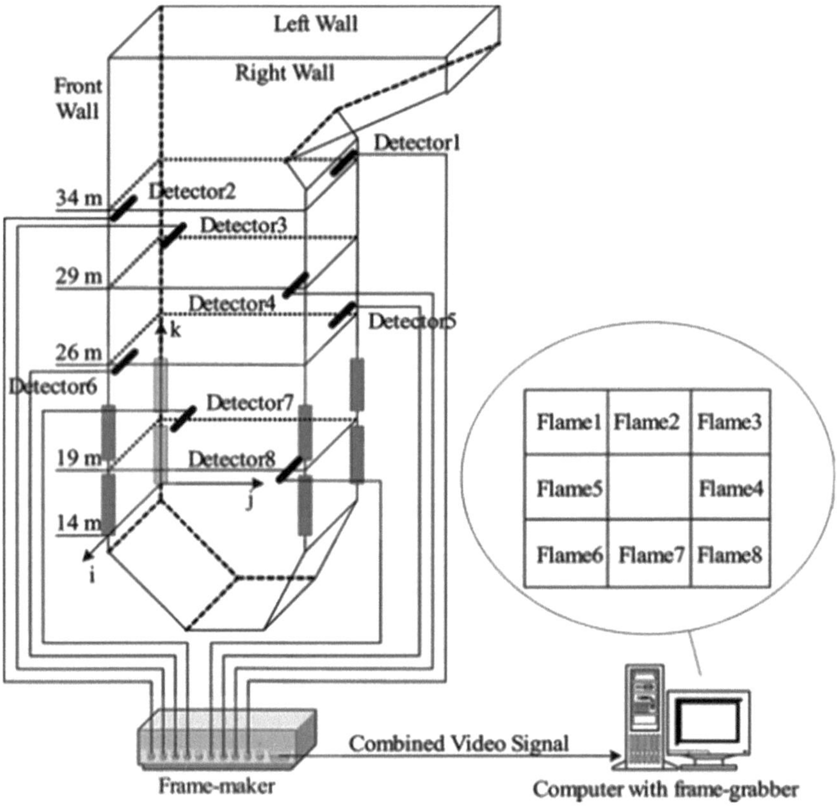

Fig. 3.10 The furnace and the experimental setup [31]

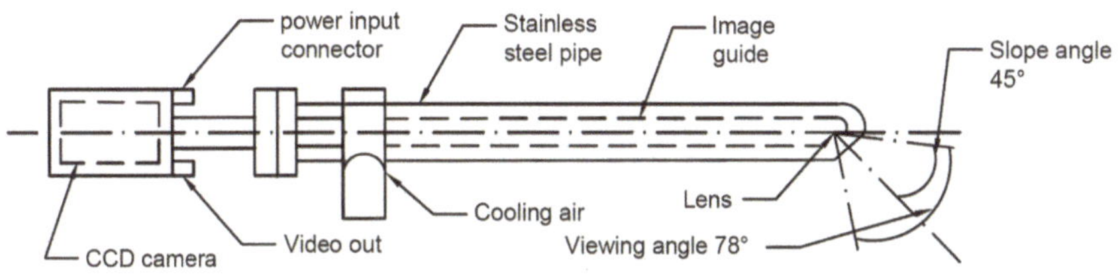

Fig. 3.11 Schematic structure of the flame image detector [31]

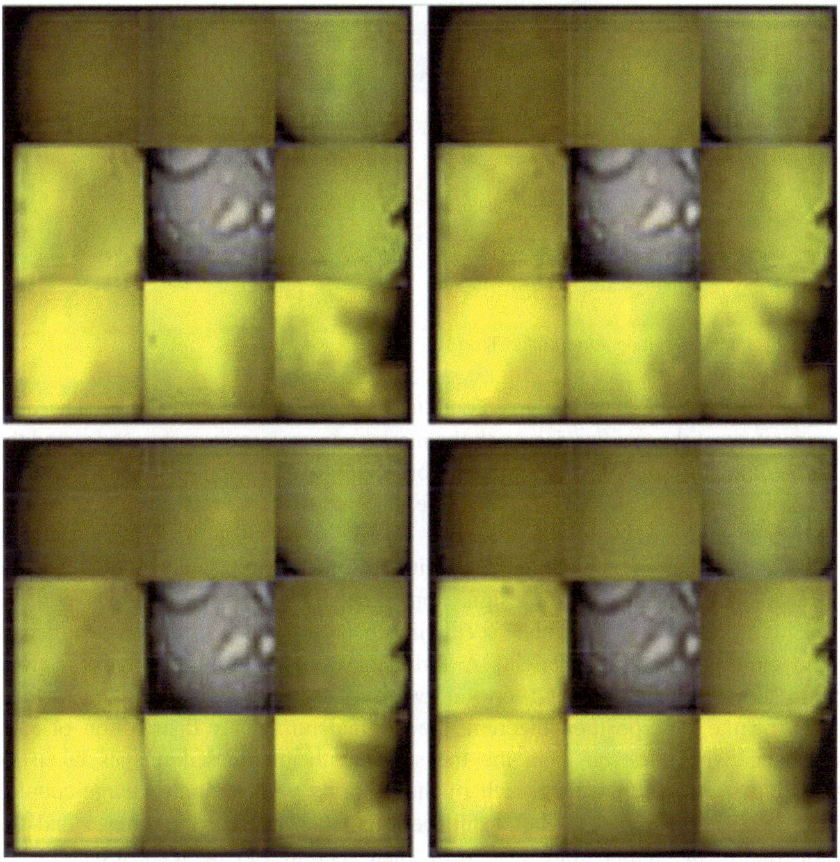

Fig. 3.12 A group of flame images [31]

(CCD) camera installed on the top of the gasifier. The reconstruction method is first applied in diesel gasification, and the reconstructed results are validated by calibrated thermocouples and a side imaging system; thus, the 3-D temperature distribution of coal-water slurry (CWS) gasification is reconstructed. The results show that the temperature distribution is more spatially homogeneous, and the average

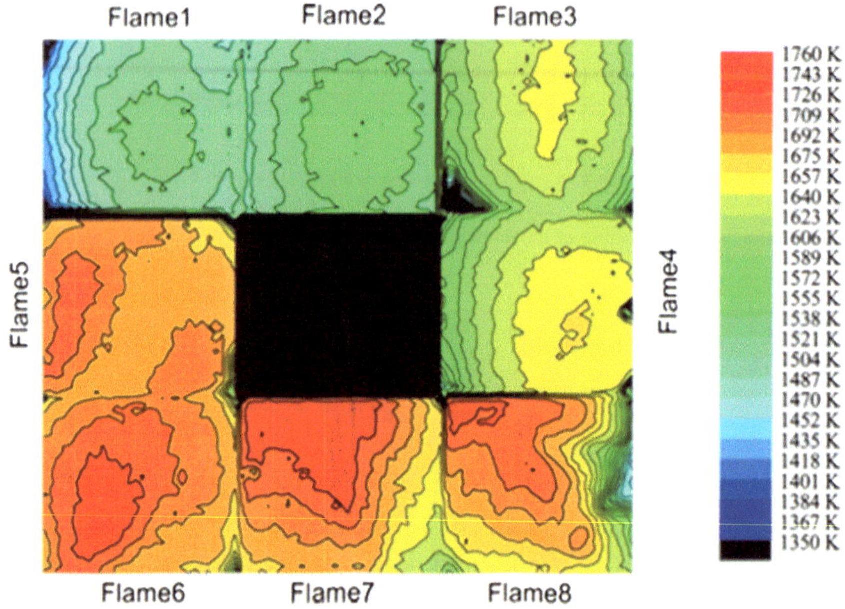

Fig. 3.13 A flame temperature image Tm [31]

temperature is higher with blurry contours of flame than that of diesel. The bulk temperature of the gasifier rises with the increase in oxygen-to-carbon ratio, and the high-temperature region which ranges from 1700 to 2200 K remains in the axis of the gasifier steadily and remains in a safe distance from the refractory wall, which makes the temperature of the refractory wall below 1550 K and makes the participating medium maintain a stable condition for gasification (Figs. 3.14 and 3.15).

Zhou [33] introduces the design, implementation, and evaluation of a flame imaging system based on optical sectioning method to reconstruct the steady-state temperature field of candlelight. A three-dimensional luminous object can be regarded as a combination of a series of two-dimensional luminous sections; hence, based on the superposition theorem, the original luminosity distribution of each section can be retrieved from the focused and the defocused images taken by adjusting the focus plane on each of the sections with a single charge-coupled device camera. From the relationship between the gray level of image and the temperature, the temperature profile of each section can be reconstructed. The experimental results prove the feasibility of the proposed method, which can be used as a new measurement and diagnostic tool for quantitative characterization of steady-state flame (Figs. 3.16 and 3.17).

Liu [34] introduces a means of CCD cameras through the inverse radiation analysis to reconstruct the two-dimensional (2-D) and three-dimensional (3-D) temperature distributions in laboratory-scale and large-scale pulverized coal-fired furnaces. The reconstruction technique was developed based on the backward Monte Carlo method. The reconstruction matrix equation was solved by the

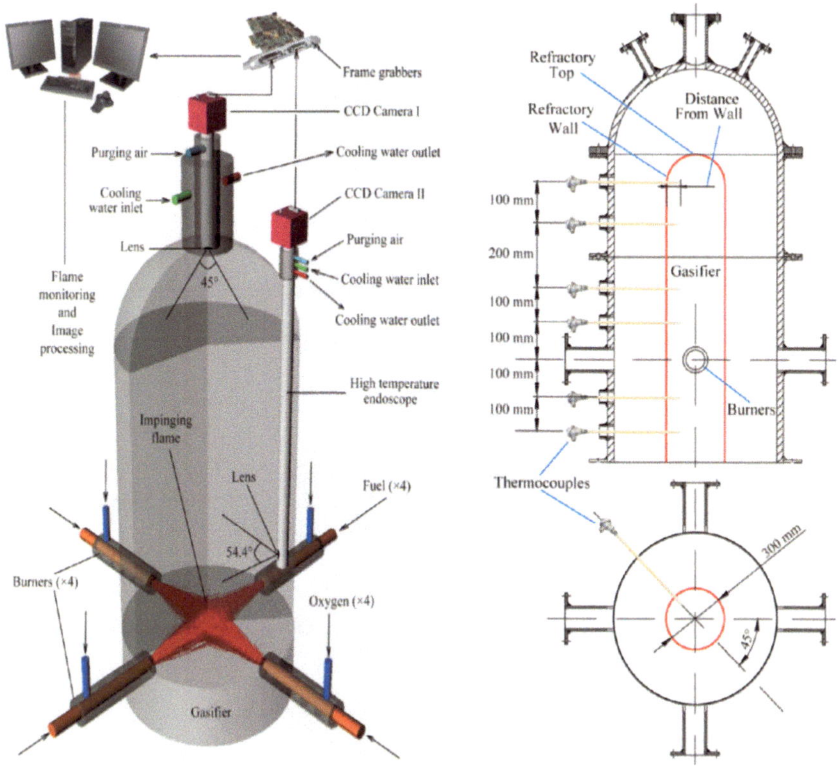

Fig. 3.14 Schematic of the bench-scale OMB gasifier; locations of thermocouples [32]

least-square QR decomposition (LSQR) method. The numerical reconstructions were utilized for checking the validity of inverse method before each experimental reconstruction. The experimental reconstruction results were compared with those in references. The temperature distributions for both laboratory-scale and large-scale pulverized coal-fired furnaces can be numerically reconstructed well, and the experimentally reconstructed temperature distributions can reasonably reproduce the main features of actual temperature distributions in the pulverized coal-fired furnaces. The reconstruction technique also has a good ability of online temperature distributions reconstruction and monitoring, and will be useful for the combustion diagnostics and control of pulverized coal-fired power plant (Figs. 3.18, 3.19, 3.20, 3.21, 3.22, 3.23, and 3.24).

Huang [35] proposes an approach of using flame emission images from four CCD camera detectors installed on the furnace wall to reconstruct 3-D flame temperature through inverse radiation analysis. The scattering from participating medium in the flame was considered by combining the discrete radiative transfer method with the discrete ordinate method. A modified minimum-residual algorithm was employed to calculate the least squares solution of the ill-conditioned inverse

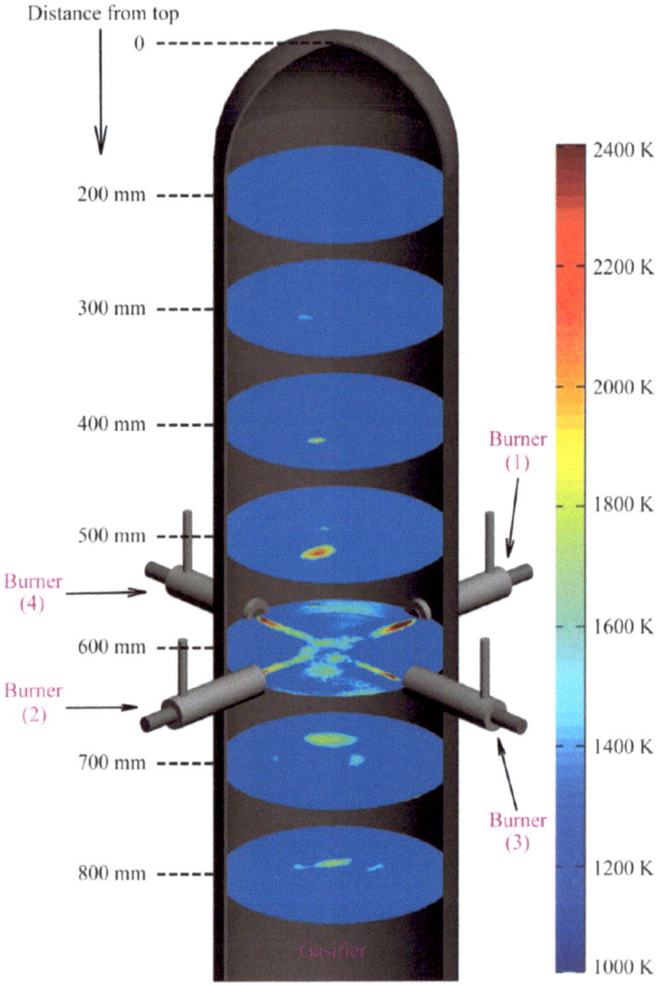

Fig. 3.15 Three-dimensional temperature distribution of diesel gasification [32]

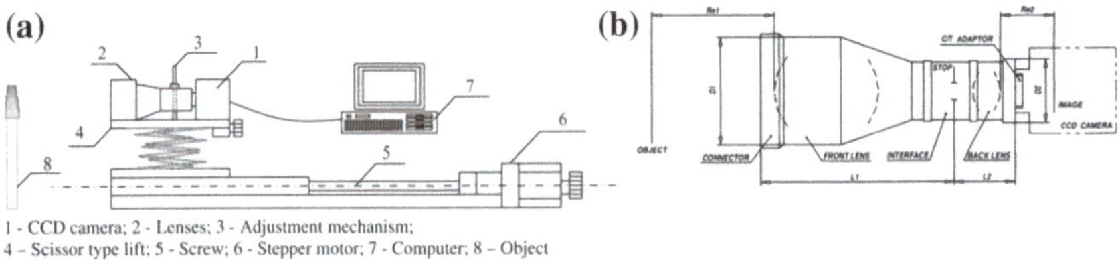

Fig. 3.16 Schematic of the experimental setup and configuration of the lenses system [33]

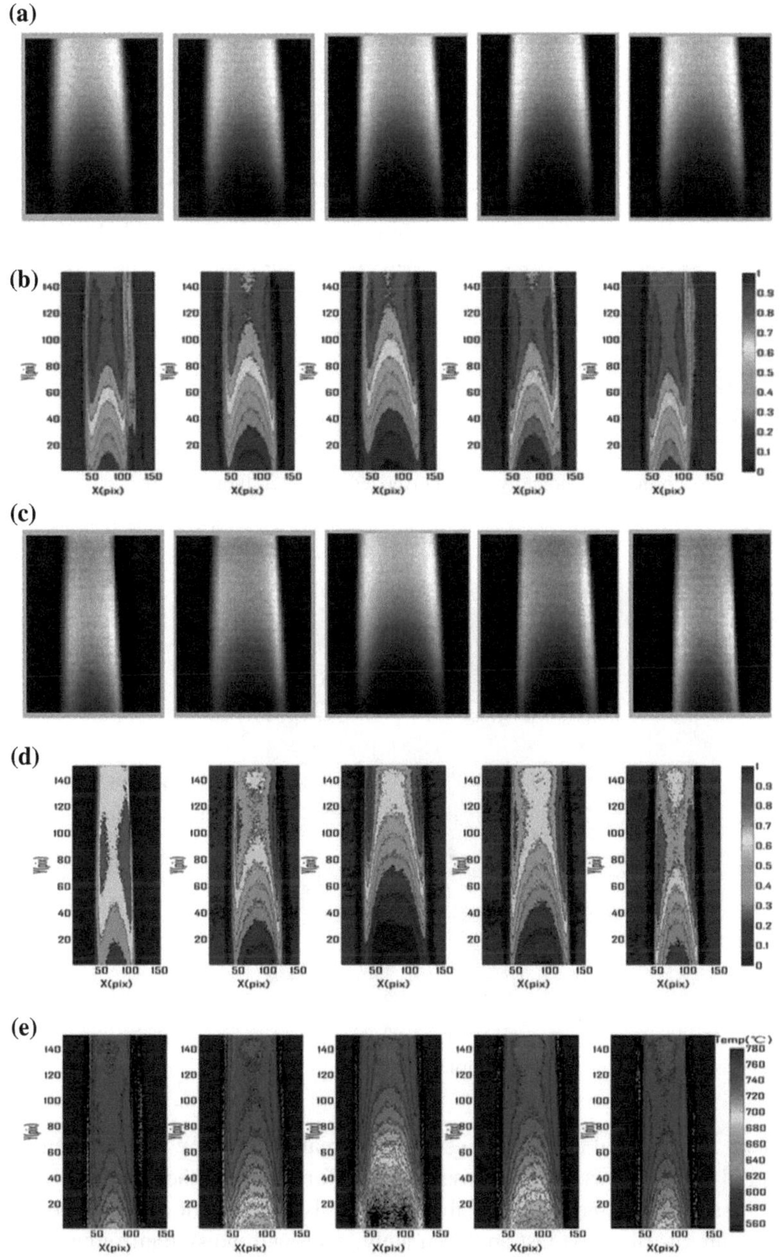

Reconstruction of lower part of the candlelight

Fig. 3.17 **a** Gray-level images captured by CCD camera for different longitudinal sections. **b** Normalized luminosity distributions for different longitudinal sections. **c** Reconstructed gray-level images for different longitudinal sections. **d** Reconstructed normalized luminosity for different longitudinal sections. **e** Reconstructed temperature distributions for different longitudinal sections [33]

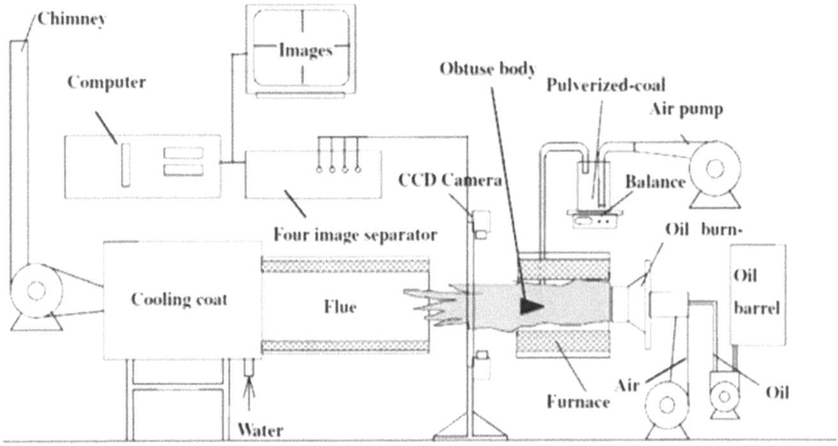

Fig. 3.18 Sketch map of laboratory-scale pulverized coal-fired furnace [34]

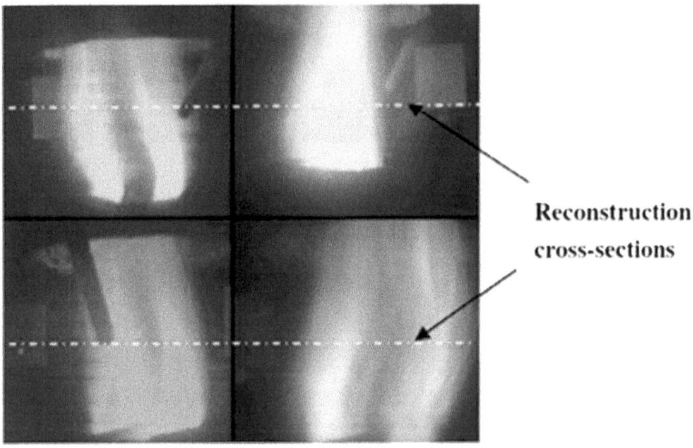

Fig. 3.19 Flame images of four CCD cameras and reconstruction cross sections for laboratory-scale pulverized coal-fired furnace [34]

problem. A numerical test problem simulating real temperature measurements in an industrial furnace was used to assess the performance of the proposed method. These assessments indicate that this method is capable of reconstructing 3-D temperature distributions fast and accurately, even with noisy flame emission data. Such a capability has potential in real-time temperature measurement for combustion optimization and pollution emission control (Figs. 3.25 and 3.26)

Liu [36] *introduces a new nonintrusive measurement technique to measure the three-dimensional (3-D) soot temperature and volume fraction fields in the axisymmetric or asymmetric flames with charge-coupled device (CCD) cameras.*

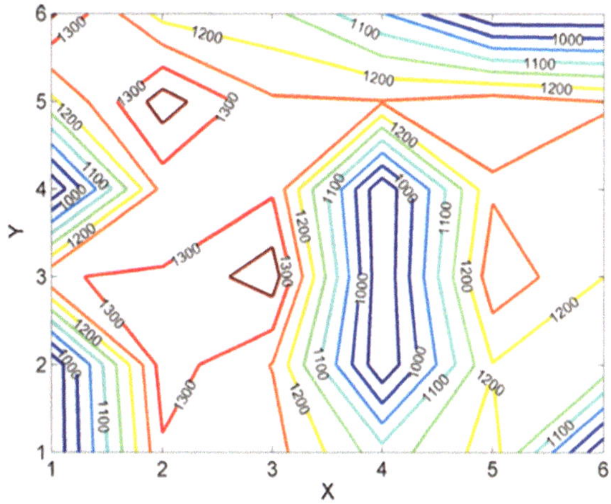

Fig. 3.20 2-D experimentally reconstructed temperature distribution of laboratory-scale pulverized coal-fired furnace [34]

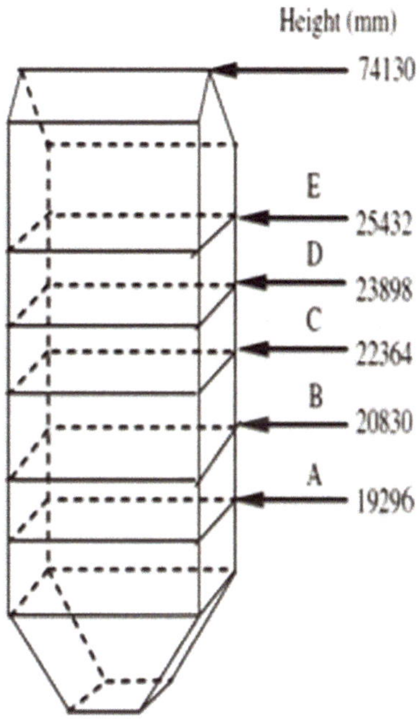

Fig. 3.21 Sketch map of 300 MW pulverized coal-fired furnace [34]

CCD cameras were introduced to capture the flame images for obtaining the line of sight radiation intensities. The distributions of local emission source under two wave lengths can be deduced through solving the reconstruction matrix equation by the least-square QR decomposition method from the knowledge of the line of sight radiation intensities of the flames. The two-color distributions of the local emission source were used to retrieve the soot temperature and volume fraction distributions. The effects of the discrete ray number of CCD cameras, the number of CCD

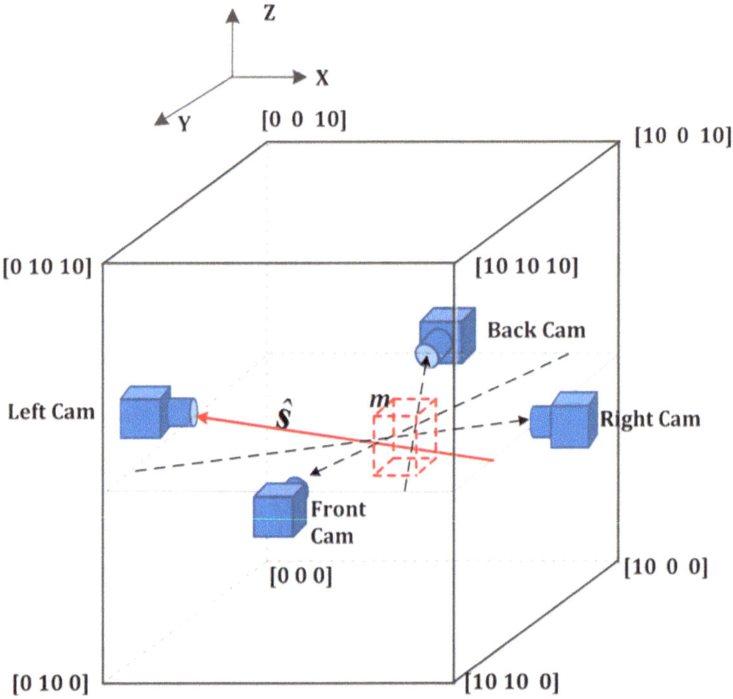

Fig. 3.25 The 3-D flame temperature reconstruction scheme for the test problem using four CCD cameras [35]

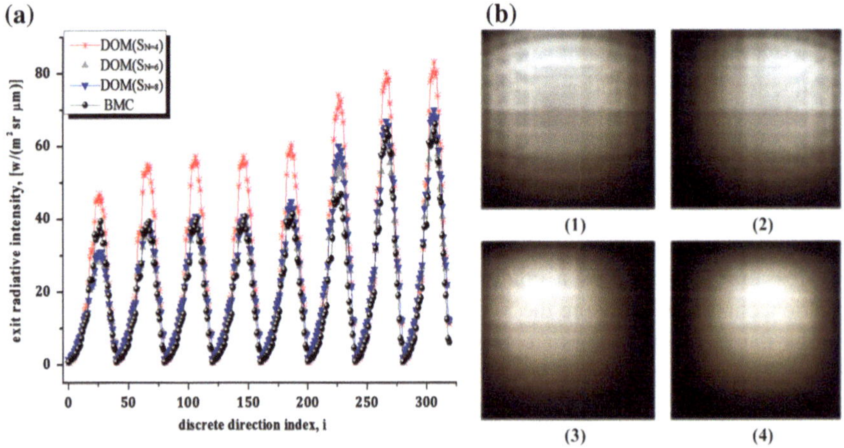

Fig. 3.26 **a** Comparison of flame emission intensities of the first 320 directions for the left camera obtained by BMC method and DOM with different ordinate sets. **b** Gray-scaled flame emission intensities for temperature reconstruction from cameras installed on (1) left, (2) right, (3) back, and (4) front walls [35]

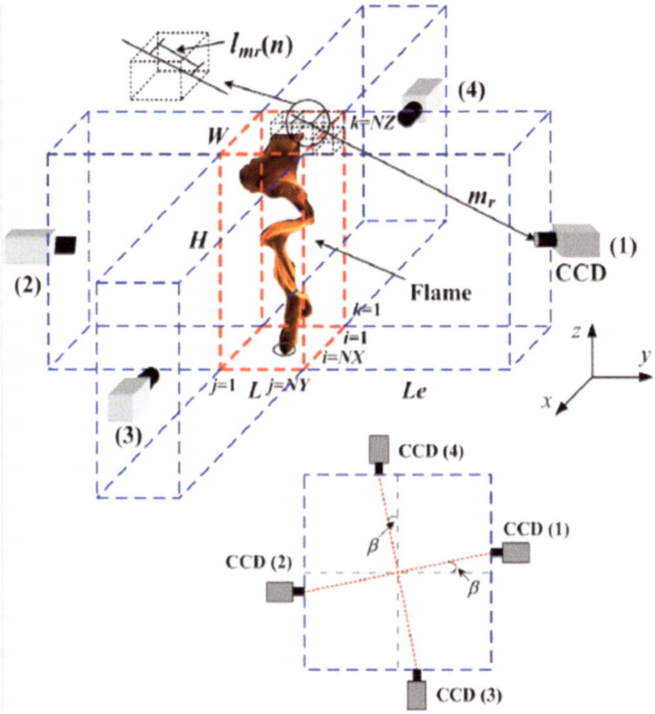

Fig. 3.27 Reconstruction system [36]

test rig are also presented and discussed. The accuracy of the temperature measurement has been verified using a precalibrated tungsten lamp, and the maximum relative error is 1.76% for eight optical fiber bundles in the temperature range from 1300 to 1550 °C. Experimental results have demonstrated that the developed technique has successfully reconstructed the 3-D temperature distribution of the flame (Figs. 3.34, 3.35, and 3.36).

Kashif [39] takes advantage of a laser extinction method to measure two-dimensional soot volume fraction field time history inside an axisymmetric diffusion flame experiencing slow unsteady phenomena preserving the symmetry. The use of a continuous wave laser as the light source enables this repetition rate. The technique is shown to allow a fine description of the soot volume fraction field in a flickering flame exhibiting a 12.6 Hz flickering phenomenon. Within this range of repetition rate, the technique and its subsequent postprocessing require neither any method for time-domain reconstruction nor any correction for energy intrusion. Possibly complemented by such a reconstruction method, the technique should support further soot volume fraction database in oscillating flames that exhibit characteristic times relevant to the current efforts in the validation of soot processes modeling (Figs. 3.37 and 3.38).

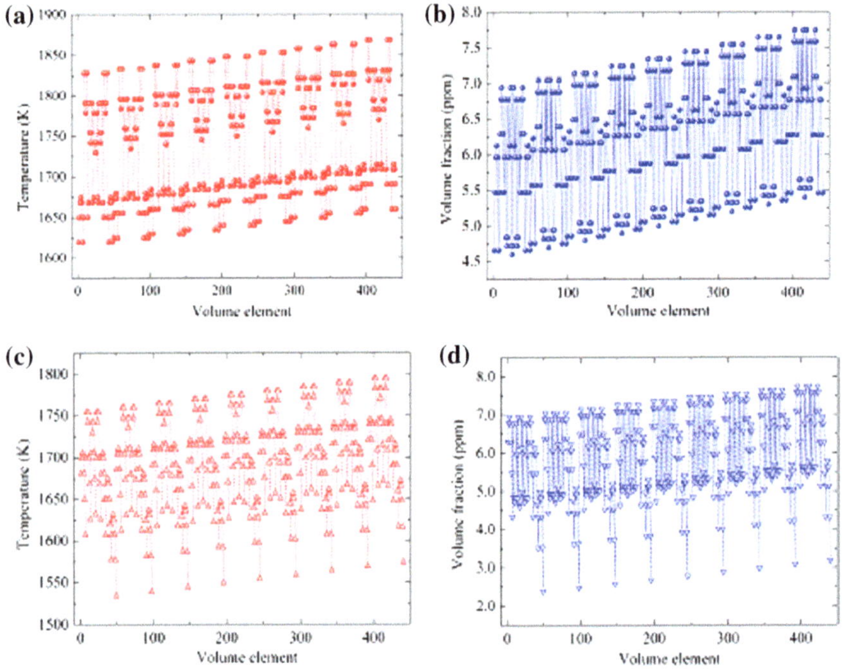

Fig. 3.28 Assumed 3-D soot temperature and volume fraction distributions: **a** axisymmetric temperature distribution; **b** axisymmetric volume fraction distribution; **c** asymmetric temperature distribution; and **d** asymmetric volume fraction distribution. Figures **a** and **b** were served as the exact ones for validation of the reconstruction method developed

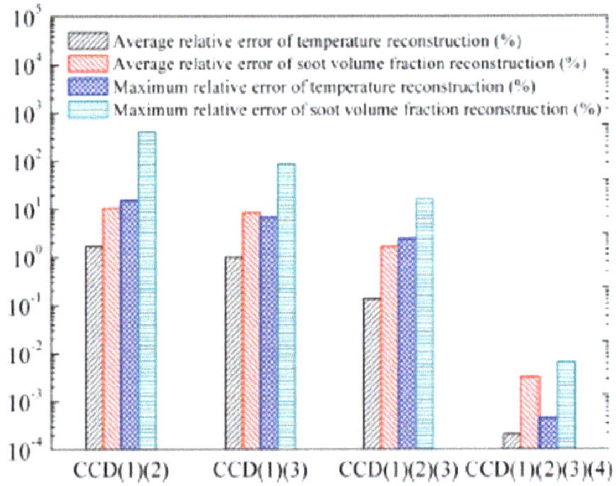

Fig. 3.29 Effects of different combinations of the CCD cameras on the reconstruction accuracy [36]

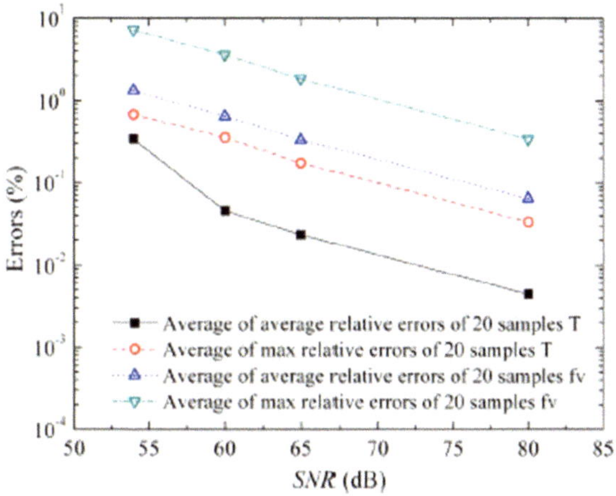

Fig. 3.30 Effects of SNR on the reconstruction accuracy [36]

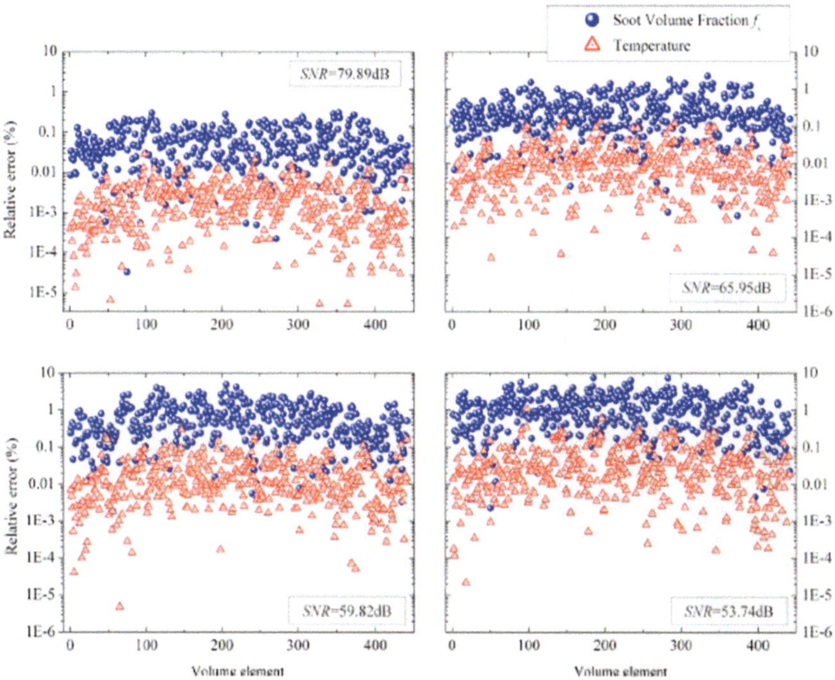

Fig. 3.31 Relative errors of reconstruction with different measurement errors [36]

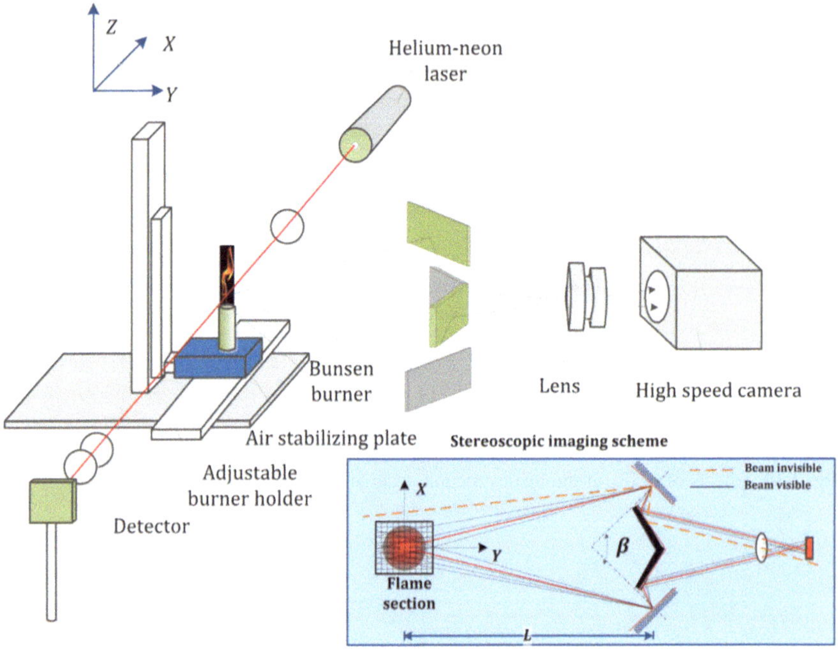

Fig. 3.32 Stereoscopic imaging system using a single high-speed camera [37]

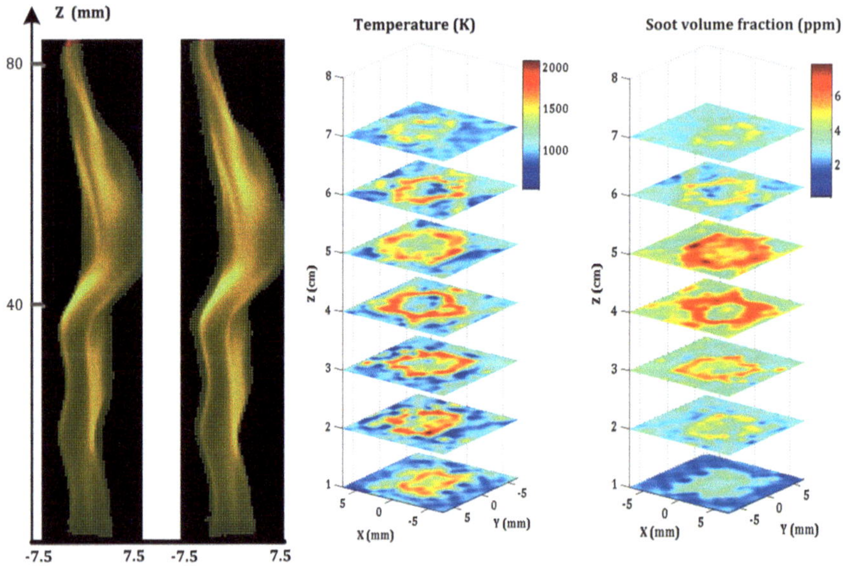

Fig. 3.33 Stereoscopic image and reconstructed temperature and soot volume fraction [37]

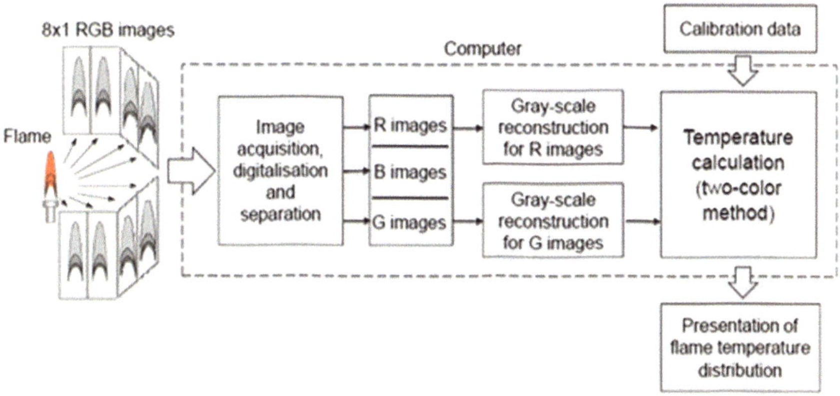

Fig. 3.34 Block diagram of 3-D reconstruction of flame temperature distribution based on optical tomographic and two-color pyrometric techniques [38]

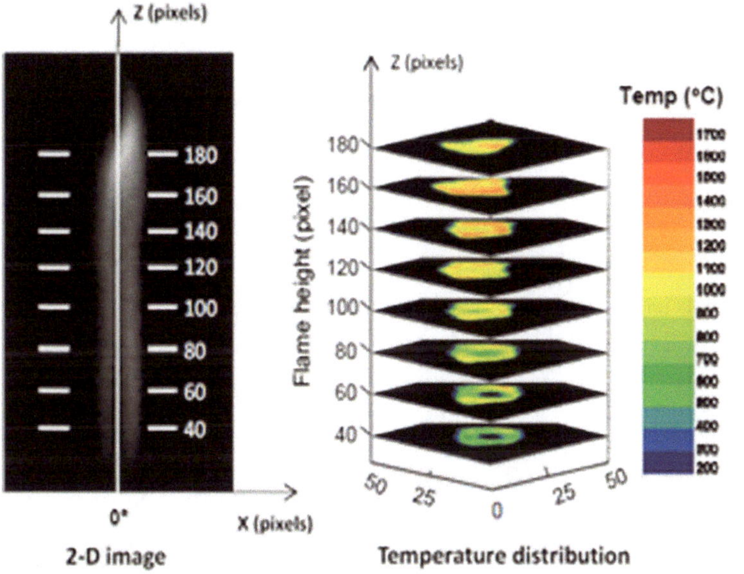

Fig. 3.35 2-D image of the flame and 3-D temperature distribution of the flame cross sections [38]

Hossain [40] presents an optical fiber imaging-based tomographic system for the 3-D visualization and characterization of a burner flame. The system consists of eight imaging fiber bundles coupled with two RGB charge-coupled device cameras which are used to acquire flame images simultaneously from eight different directions around the burner. The fiber bundle has 30k picture elements and an objective lens with a 92° angle of view. The characteristic evaluation of the imaging

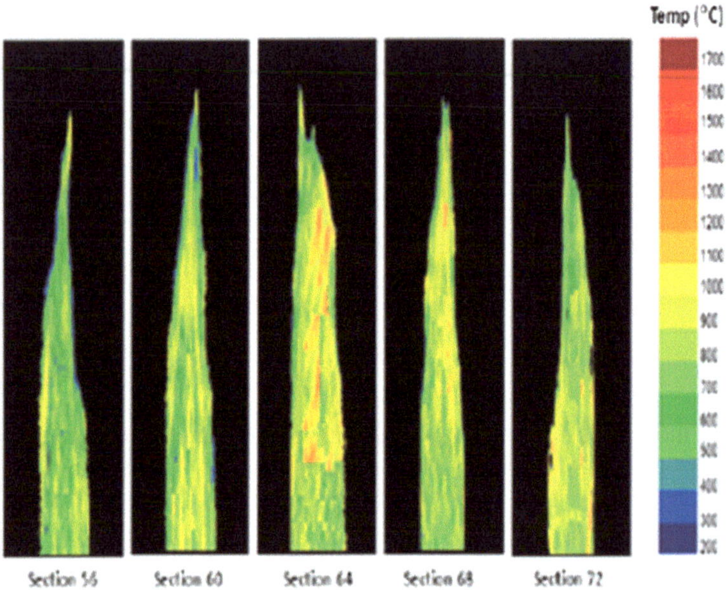

Fig. 3.36 3-D temperature distribution of longitudinal sections [38]

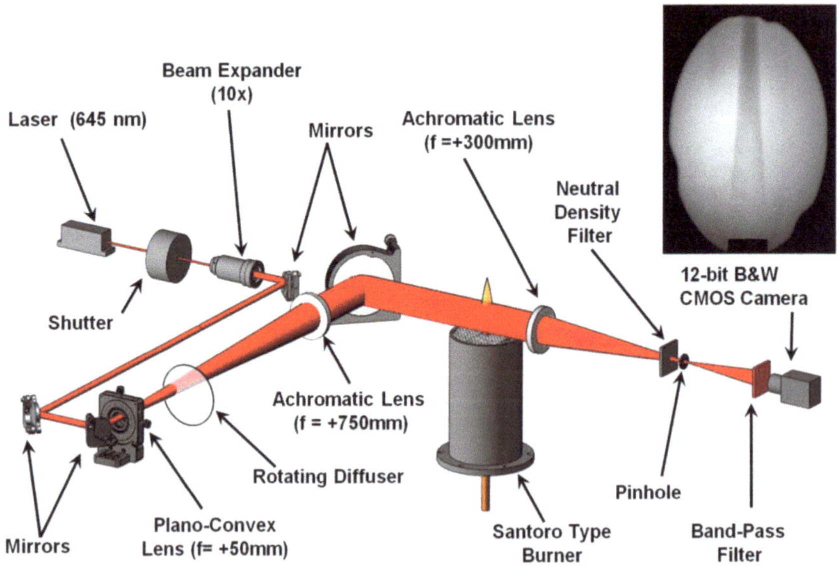

Fig. 3.37 Schematic of the optical arrangement for the 2-D LEM technique. The inset shows a typical frame when the shutter is open and the flame is established [39]

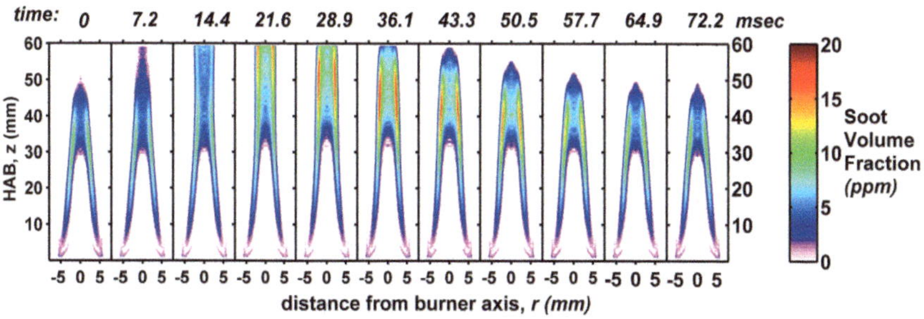

Fig. 3.38 Reconstructed time history of soot volume fraction fields in the flickering flame [ethylene flow rate = 6 ± 0.12 cm^3/s, oxidizer flow rate = 1250 ± 33.33 cm^3/s, oxygen mole fraction in oxidizer stream (XO_2: 0.55), balance gas: CO_2] [39]

fiber bundles and the calibration of the system were conducted to ensure the accuracy of the system. A new tomographic algorithm that combines the logical filtered back-projection and the simultaneous algebraic reconstruction technique is proposed to reconstruct the flame sections from the images. A direct comparison between the proposed algorithm and other tomographic approaches is conducted through computer simulation for different test templates and numbers of projections. The effectiveness of the imaging system and computer algorithm is assessed through experimental tests (Figs. 3.39, 3.40, and 3.41).

Anikin [41] makes use of a new developed fast tomographic reconstruction device to detect 2-D chemiluminescence distributions of OH* in reaction zones of a near laminar and a turbulent diffusion flame. The emission of OH*-chemiluminescence originating from the reaction zones of the flame fronts was registered by ten Kepler telescopes surrounding the object under investigation at different predefined angles. The signals emerging from each telescope are collected by a fiber

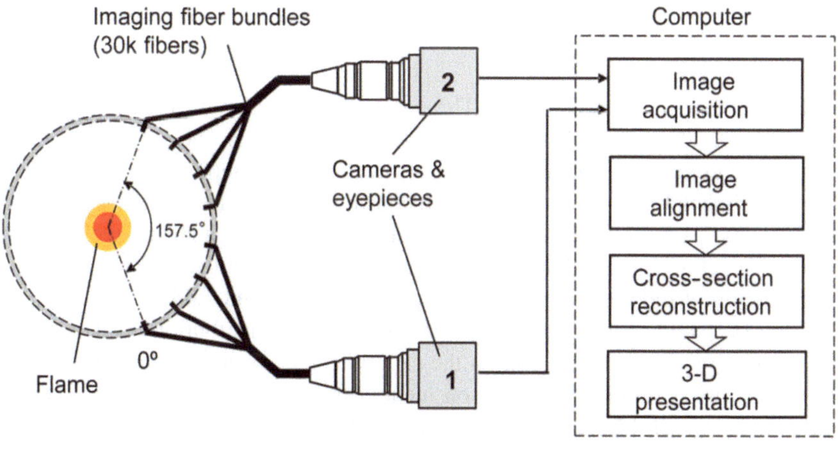

Fig. 3.39 Schematic of the 3-D flame imaging system [40]

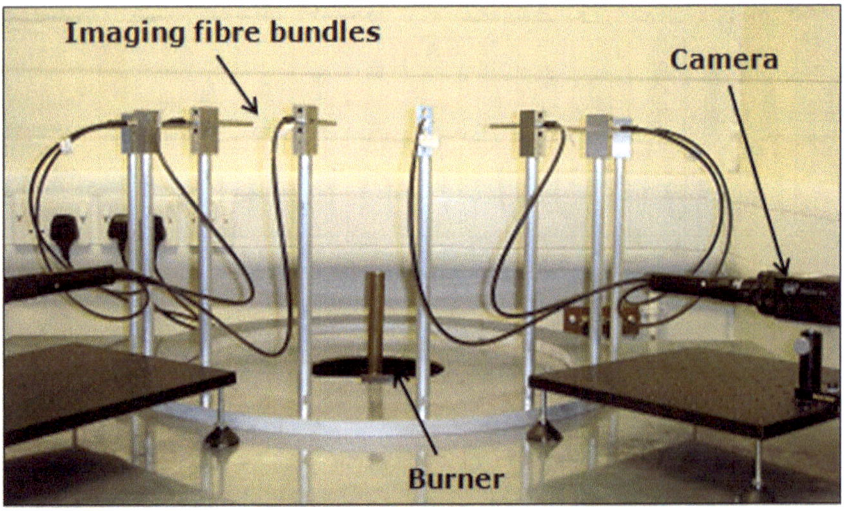

Fig. 3.40 Physical implementation of the 3-D flame imaging system [40]

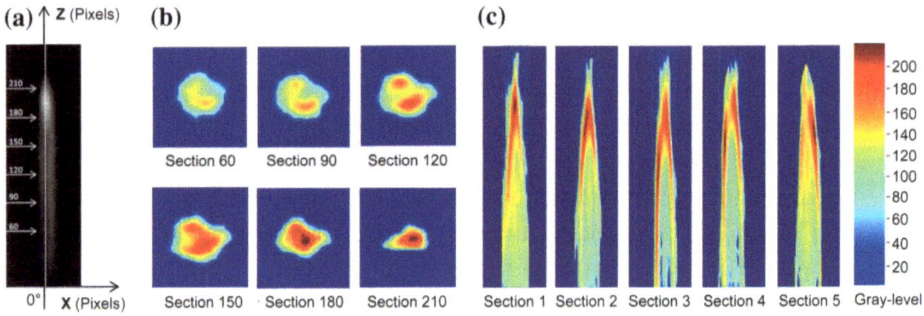

Fig. 3.41 Luminosity reconstruction of flame cross and longitudinal sections. **a** Two-dimensional image. **b** Cross sections. **c** Longitudinal sections [40]

cable consisting of 90 single fibers. These signals are imaged by a relay-optics onto the photocathode of a single image intensified CCD camera. The output data of the camera are used for the reconstructions of the 2-D distributions of OH*-emission. The experimental results show that the reconstructions obtained at exposure times down to 200 μs reproduce fine structures of the flames with a spatial resolution of 1 mm. Therefore, the method used in this paper is a useful tool for the detailed investigation of turbulent combustion (Figs. 3.42, 3.43, and 3.44).

Chen [42] studies the effect of species composition distribution on flow field's temperature reconstruction based on the experimentally measured refractive index. A temperature partition reconstruction method based on flow field's phase distribution is proposed. Moiré deflection tomography is chosen to obtain the refractive index of a flame with probe wavelength of 808 nm. Enlightened by the visual

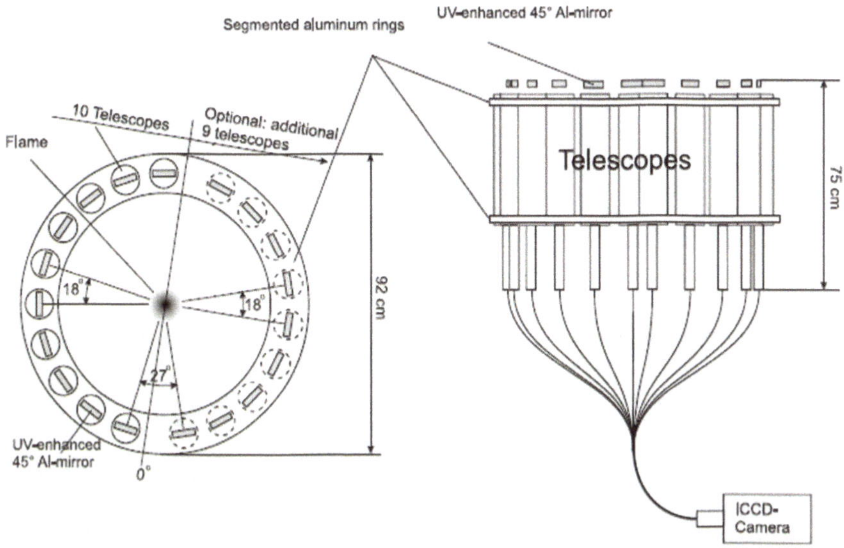

Fig. 3.42 Schematic illustration of the tomographic reconstruction setup. Left figure: top view, right figure: side view of the arrangement consisting of 10 Kepler telescopes (optional 19). The OH*-chemiluminescence signals propagating in the horizontal plane are deflected into the vertical direction by UV-enhanced 45 mirrors [41]

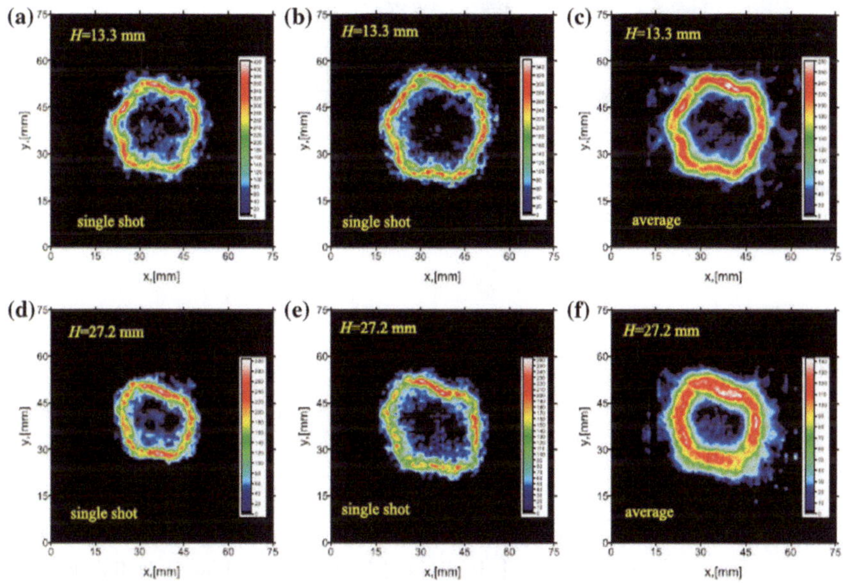

Fig. 3.43 The reconstructed distributions of the OH*-chemiluminescence emission of the nearly laminar (cylindrical) flame [41]

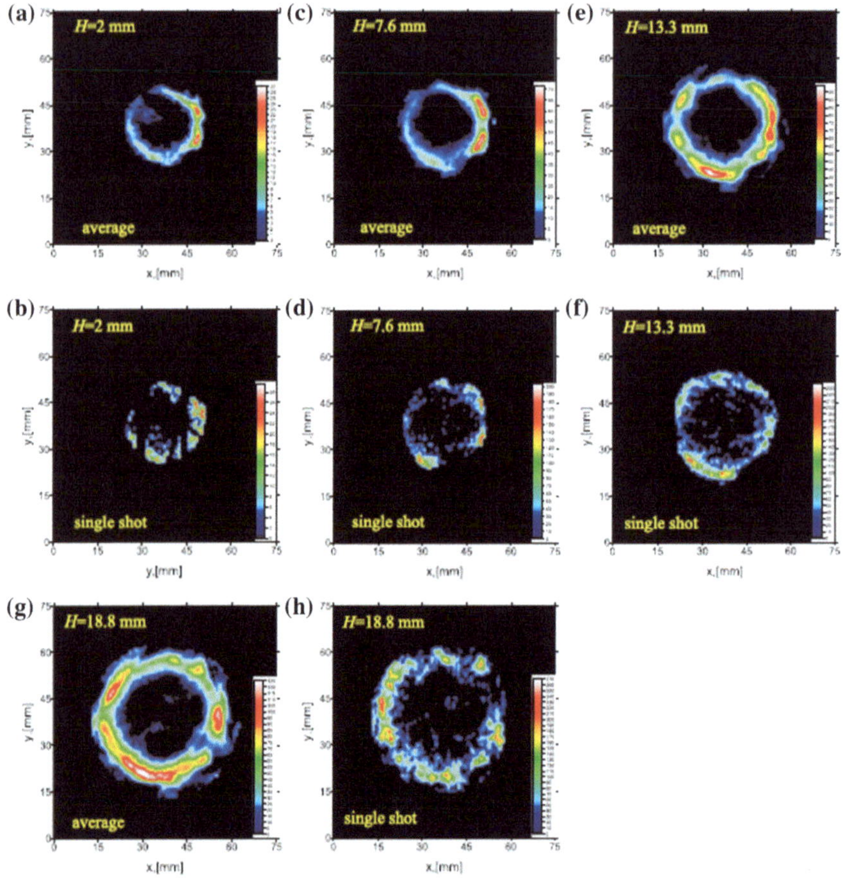

Fig. 3.44 The reconstructed distributions of the OH*-emission of the turbulent (conical) flame are represented in the horizontal planes [41]

Fig. 3.45 The schematic diagram of moiré deflectometry

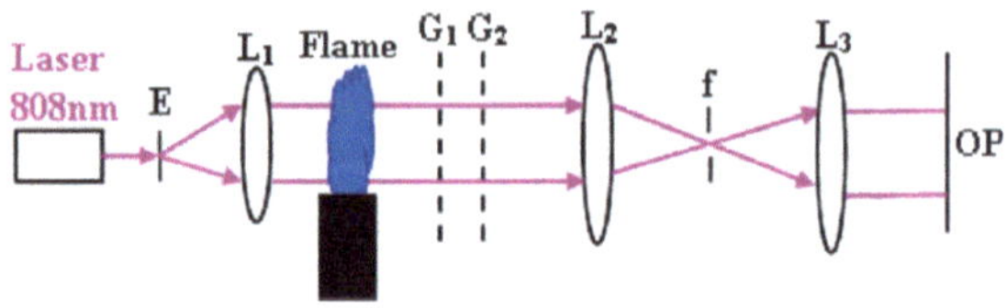

characteristic of optical computerized tomography (OCT) for flow field and the practical structure of the flame, it is divided into three regions in radial direction based on its phase distribution. The temperature is reconstructed with adopting different species composition for different regions. Finally, the rationality of the method is analyzed (Figs. 3.45, 3.46, 3.47, and 3.48).

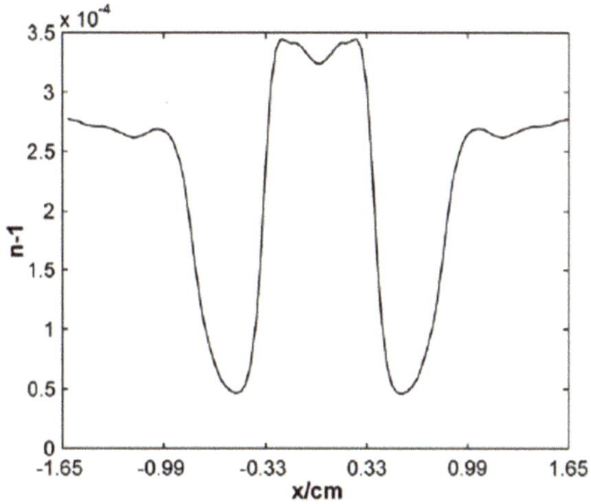

Fig. 3.46 The radial refractive index distribution [42]

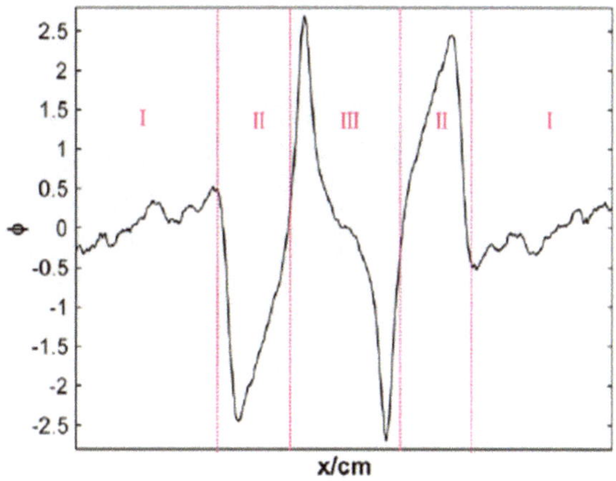

Fig. 3.47 The divided regions in radial direction [42]

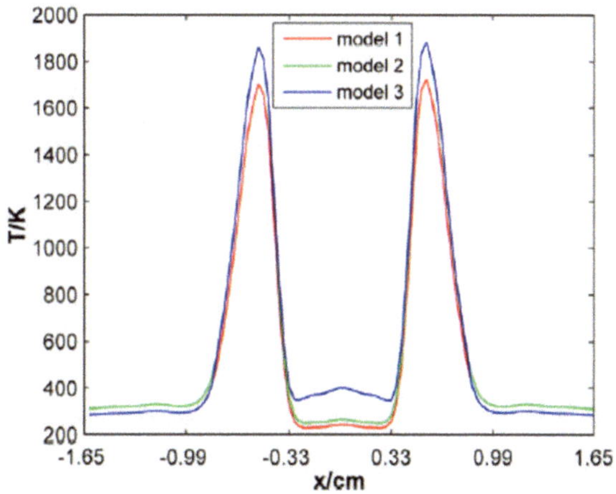

Fig. 3.48 The radial temperature distributions with different models [42]

Ishino [43] builds a test bench with a forty-lens camera around it. The advanced CT (computerized tomography) reconstruction technique is presented to measure an instantaneous three-dimensional distribution of chemiluminescence of a turbulent premixed flame. First, instantaneous two-dimensional images of an objective flame are simultaneously taken from forty horizontal directions with a forty-lens camera. Then, four hundred horizontal CT images, which are reconstructed from the "projection" images by maximum likelihood expectation maximization (MLEM) algorithm, are vertically accumulated. Finally, an instantaneous three-dimensional distribution of flame chemiluminescence is obtained. However, the amount of data transmission is very large because of the forty-lens results in a long reconstruction process and a poor real time. Various types of display of three-dimensional distribution are also performed to demonstrate the three-dimensionality of data acquisition (Figs. 3.49, 3.50, 3.51, 3.52, and 3.53).

Fig. 3.49 Appearance (left) and close-up (right) of multi-lens camera which equipped with forty small lenses [43]

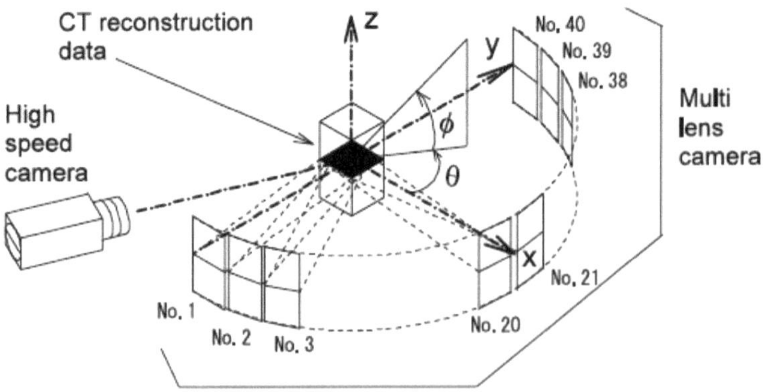

Fig. 3.50 Schematic diagram of the experimental apparatus and coordinate system [43]

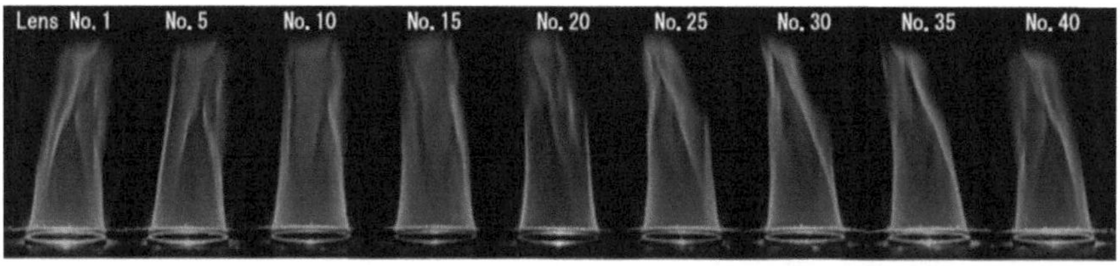

Fig. 3.51 Sample set of "projection" images taken simultaneously by multi-lens camera [43]

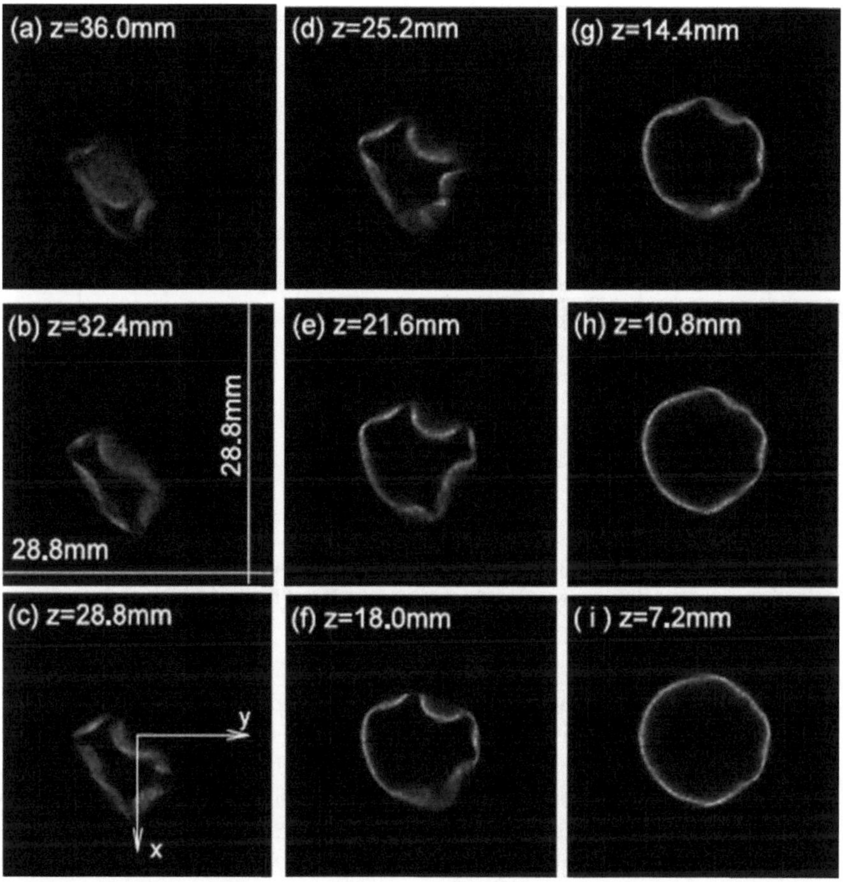

Fig. 3.52 Samples of horizontal reconstruction of turbulent premixed flame of propane–air rich mixture [43]

Lammel [44] introduces an EMS infrared spectrometer for selective and quantitative chemical gas analysis. Infrared absorption spectroscopy can distinguish gases easily and also detect nonreactive molecules like CO_2. A tunable interference filter

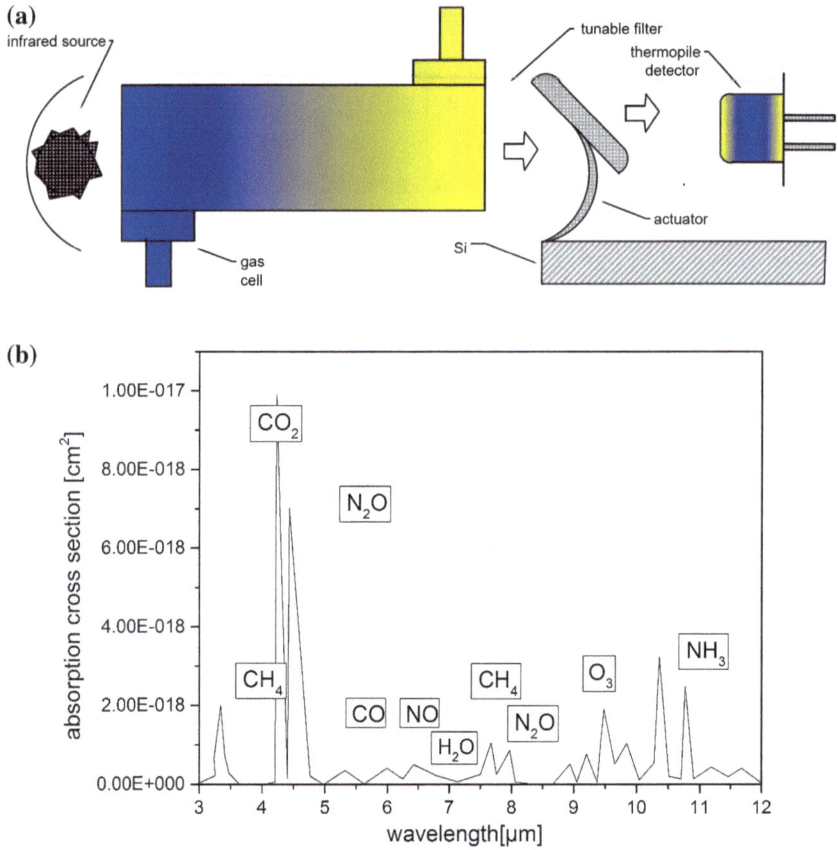

Fig. 3.53 **a** Operating principle of the MEMS infrared gas spectrometer. It measures the infrared absorption of a gas at different wavelengths that can be selected by a tunable interference filter, as shown in Fig. (**b**) [44]

scans the desired part of the infrared spectrum. A single pixel thermopile detector measures serially the intensity at selected wavelengths. The tunable optical interference filter is fabricated by a new porous silicon batch technology using only two photolithography steps. The refractive index of this filter microplate is gradually modulated in depth to create a Bragg mirror, edge filter, or a Fabry–Perot band-pass filter for central wavelengths between 400 nm and 8 p. Two thermal bimorph micro-actuators tilt the plate by up to 90°, changing the incidence angle of the beam to be analyzed. This tunes the wavelength transmitted to the detector. The filter area can be chosen between 0.27 mm × 0.70 mm and 2.50 mm × 3.00 mm, and its thickness is typically 30 p. Results showed that CO_2 and CO can be detected selectively with this system—which is interesting in combustion processes—by measuring their absorption at 4.26 μm and 4.65 pm, respectively (Figs. 3.54 and 3.55).

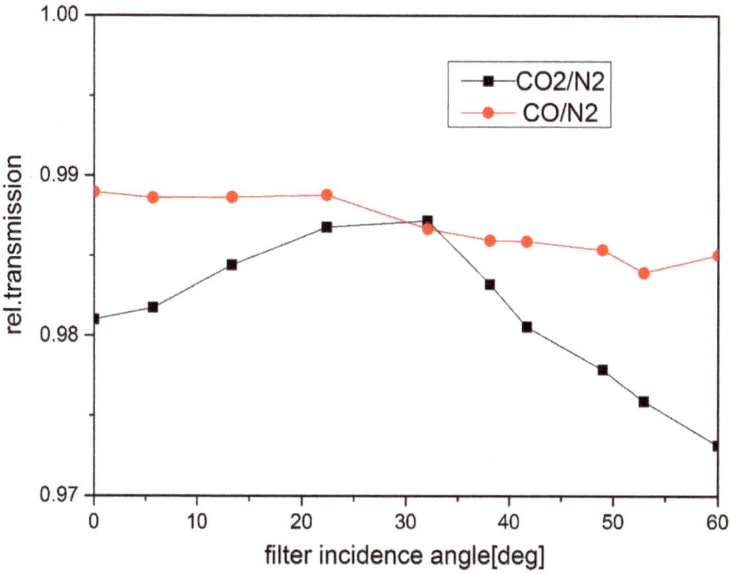

Fig. 3.54 Measured overall transmission (gas cell and filter) for different gases relative to a N_2 reference, using the setup of Fig. 3.53a. The incidence angle of the filter was scanned from 0° to 60° corresponding to a wavelength sweep like CO_2 (0.5%) and CO (1.0%) can be clearly separated by their different infrared fingerprints [44]

3.3 CFD Method

In situ tests in real furnace are the most reliable way to get data. Unfortunately, these tests demand more time, labor, and money with the increasing in boiler capacity. The use of modeling method is a compromising but quite not good option because it is still very expensive to make a test rig for thermal simulation. What's worse, the accuracy of amplified modeling results is hardly confident for the complexity of turbulence, reaction flow and combustion, and pollutant formation. Comprehensive numerical simulation in the furnace can simulate complicated flow, combustion, and pollutant formation in any scale. Due to its convenient variation of operating parameters and conditions, comprehensive information for designers and operators, and lower cost of money and time than modeling method, comprehensive numerical simulation in the furnace is gradually becoming an effective way for boiler designing, improvement, and operation optimization.

In pulverized coal boiler, for example, the main framework of CFD (computational fluid dynamics) method for comprehensive numerical simulation in the furnace is to solve three-dimensional conservative partial equations for mass, energy, and momentum, and some other submodels coupled with main framework to obtain the mixture and reactions of gas-phase species, coal devolatilization, char combustion, radiative heat transfer, and so on.

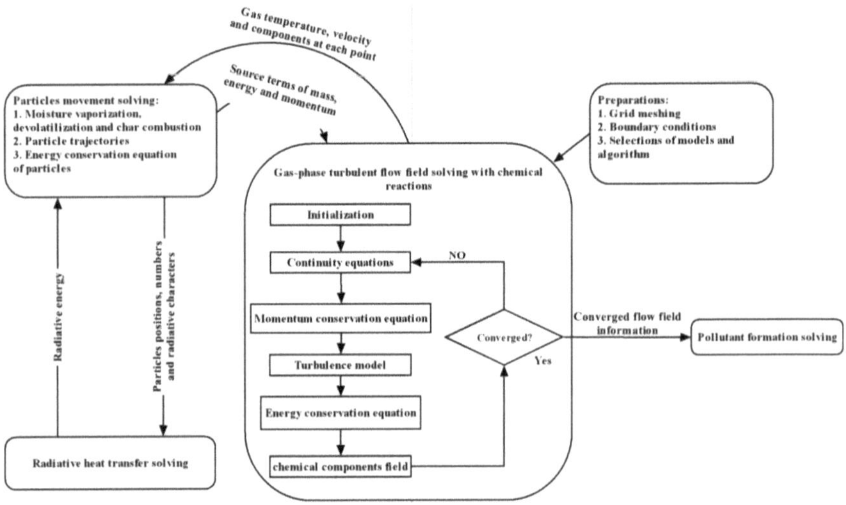

Fig. 3.55 Solution procedures for comprehensive numerical simulation in the furnace

Following information can be obtained through comprehensive numerical simulation in the furnace: temperature field, velocity field, distribution of gas species, particle trajectory, particle burnout level, particle density field, ash accumulation, and so on. Besides all the data we can almost get from in situ tests, the simulation can provide the data we cannot get on site. A large number of comprehensive combustion simulators have been developed. Some commercial software such as FLUENT, CFX, PHEONICS, and STAR-CD are capable of comprehensively simulating combustion in the furnace. Differences among these software are mainly in selecting models.

Figure 3.56 clearly shows solution procedures for comprehensive numerical simulation and relationships between main framework and submodels as a general concept for pulverized coal combustion simulation.

3.3.1 Turbulence Model

Turbulence research is the basis of combustion issues in the boiler. Its accuracy level greatly influences subsequent simulations of combustion, pollutant formation, de-NO_x process, discrete phase, and slagging. Current research methods for turbulence problem consist of DNS (direct numerical simulation), LES (large-eddy simulation), and RANS (Reynolds-averaged Navier–Stokes).

Direct numerical simulation is a method which directly solves turbulence governing equations without any modeling assumption and acquires precise description of the flow field. However, DNS of flow must be done at all temporal and spatial scales due to its own feature, which costs extremely high arithmetic speed and

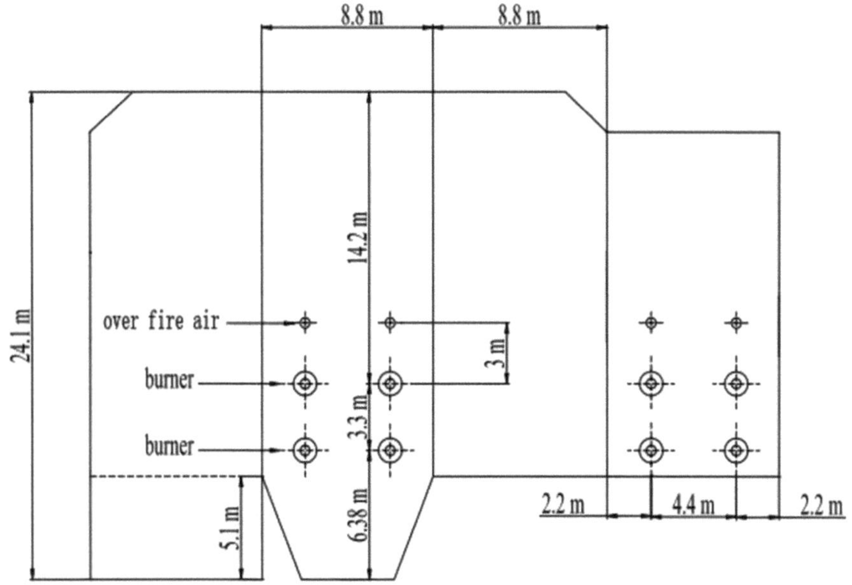

Fig. 3.56 The structure and simulation zone of half furnace [47]

memory capacity. So, DNS is only for solving laminar or low-Reynolds-number flow issues up to now. There is still a great distance to introducing DNS in industrial application where Reynolds number is very high in general. Presently, DNS is suitable to investigate turbulence mechanism in laboratory.

Large-eddy simulation is a method similar to direct numerical simulation but more practical than the latter. It filters governing equations with a selected filter width between large scale and Kolmogorov scale, and then divides all variables into large-scale ones and small-scale ones. Large-scale variables are directly solved using DNS while small-scale variables modeled using a subgrid model. Consequently, true structural states are obtained for large-scale variables. Given that small-scale eddies in turbulence is isotropic, the adoption of subgrid model is reasonable. LES has been widely used and considered to be potential for practical engineering application because of its accuracy and affordable amount of computations compared with DNS. There have been cases of single-swirl burner with LES but still there are no cases of full-scale boiler with LES.

Reynolds-averaged Navier–Stokes method has been widely used in practical engineering application. It time-averages transient governing equations and makes them containing unknown variables including time-averaged physical parameters and time-averaged product of fluctuating physical parameters. Then, all these modified time-averaged equations are closed and solved with an assumption called turbulence model. As a result, the validation of turbulence model determines the computational accuracy. Unfortunately, no single turbulence model is suitable for all issues. The choice of turbulence model will depend on considerations such as the physics encompassed in the flow, the established practice for a specific class of problem, the level of accuracy required, the available computational resources, and the amount of time available for the simulation. To make the most appropriate choice of model for your application, you need to understand the capabilities and limitations of the various options.

There are a lot of available turbulence models such as Reynolds stress model, zero-equation model, one-equation model, and two-equation models. Standard $k-\varepsilon$ model, RNG (renormalization group) $k-\varepsilon$ model, and realizable $k-\varepsilon$ model are often used in the numerical simulation of full-scale pulverized coal boiler and single-swirl burner. Here k represents turbulence kinetic energy, and ε is its dissipation rate. The general governing equation of $k-\varepsilon$ model for 3-D cases is

$$\frac{\partial(\rho u\phi)}{\partial x}+\frac{\partial(\rho v\phi)}{\partial y}+\frac{\partial(\rho \omega\phi)}{\partial z}=\frac{\partial}{\partial x}\left(\Gamma\frac{\partial\phi}{\partial x}\right)+\frac{\partial}{\partial y}\left(\Gamma\frac{\partial\phi}{\partial y}\right)+\frac{\partial}{\partial z}\left(\Gamma\frac{\partial\phi}{\partial z}\right)+S, \quad (3.1)$$

where Γ is the generalized diffusion coefficient, and S is the source term. Γ, S of each equation in the standard $k-\varepsilon$ model are given in Table 3.4. Three coefficients (c_1, c_2, c_μ) and three constants $(\sigma_k, \sigma_e, \sigma_T)$ are introduced into this equation set.

Each value of the six empirical constants in the standard $k-\varepsilon$ model is given in Table 3.5.

The RNG $k-\varepsilon$ model is similar in form to the standard $k-\varepsilon$ model. The differences are the values of the five coefficients as shown in Table 3.6. The RNG model is more accurate and reliable for a wider class of flows than the standard $k-\varepsilon$ model.

Table 3.4 Description of each term in standard $k-\varepsilon$ model

	Γ	S
u	$\eta_{\text{eff}}=\eta+\eta_t$	$-\frac{\partial p}{\partial x}+\frac{\partial}{\partial x}\left(\eta_{\text{eff}}\frac{\partial u}{\partial x}\right)+\frac{\partial}{\partial y}\left(\eta_{\text{eff}}\frac{\partial v}{\partial x}\right)+\frac{\partial}{\partial z}\left(\eta_{\text{eff}}\frac{\partial \omega}{\partial x}\right)$
v	$\eta_{\text{eff}}=\eta+\eta_t$	$-\frac{\partial p}{\partial y}+\frac{\partial}{\partial x}\left(\eta_{\text{eff}}\frac{\partial u}{\partial y}\right)+\frac{\partial}{\partial y}\left(\eta_{\text{eff}}\frac{\partial v}{\partial y}\right)+\frac{\partial}{\partial z}\left(\eta_{\text{eff}}\frac{\partial \omega}{\partial y}\right)$
ω	$\eta_{\text{eff}}=\eta+\eta_t$	$-\frac{\partial p}{\partial z}+\frac{\partial}{\partial x}\left(\eta_{\text{eff}}\frac{\partial u}{\partial z}\right)+\frac{\partial}{\partial y}\left(\eta_{\text{eff}}\frac{\partial v}{\partial z}\right)+\frac{\partial}{\partial z}\left(\eta_{\text{eff}}\frac{\partial \omega}{\partial z}\right)$
k	$\eta+\frac{\eta_t}{\sigma_k}$	$\rho G_k-\rho\varepsilon$
ε	$\eta+\frac{\eta_t}{\sigma_\varepsilon}$	$\frac{\varepsilon}{k}(c_1\rho G_k-c_2\rho\varepsilon)$
T	$\frac{\eta}{\Pr}+\frac{\eta_t}{\sigma_T}$	According to specific problem

$\eta_t=c_\mu\rho k^2/\varepsilon$

$$G_k=\frac{\eta_t}{\rho}\left\{2\left[\left(\frac{\partial u}{\partial x}\right)^2+\left(\frac{\partial v}{\partial y}\right)^2+\left(\frac{\partial \omega}{\partial z}\right)^2\right]+\left(\frac{\partial u}{\partial y}+\frac{\partial v}{\partial x}\right)^2+\left(\frac{\partial u}{\partial z}+\frac{\partial \omega}{\partial x}\right)^2+\left(\frac{\partial v}{\partial z}+\frac{\partial \omega}{\partial y}\right)^2\right\}$$

Table 3.5 Coefficient value in standard $k-\varepsilon$ model

c_μ	c_1	c_2	σ_k	σ_ε	σ_T
0.09	1.44	1.92	1.0	1.3	0.9–1.0

Table 3.6 Coefficient value in RNG $k - \varepsilon$ model

c_μ	0.085
c_1	$1.42 - \frac{\tilde{\eta}(1-\tilde{\eta}/\tilde{\eta}_0)}{1+\beta\tilde{\eta}^3}$
c_2	1.68
σ_k	0.7179
σ_ε	0.7179
where $\tilde{\eta} = Sk/\varepsilon$, $S = \left(2S_{i,j}S_{i,j}\right)^{1/2}$, $\tilde{\eta}_0 = 4.38$, $\beta = 0.015$, $S_{i,j} = \frac{1}{2}\left(\frac{\partial u_i}{\partial x_j} + \frac{\partial u_j}{\partial x_i}\right)$	

Table 3.7 Coefficient value in realizable $k - \varepsilon$ model

c_μ	$\frac{1}{A_0 + A_s U^* \frac{k}{\varepsilon}}$
where $A_0 = 4.0$, $A_s = \sqrt{6}\cos\phi$, $\phi = \frac{1}{3}\cos^{-1}\left(\sqrt{6}\,W\right)$, $W = \frac{S_{i,j}S_{j,k}S_{i,j}}{\left(S_{i,j}S_{i,j}\right)^{3/2}}$, $U^* = \sqrt{S_{i,j}S_{i,j} + \tilde{\Omega}_{i,j}\tilde{\Omega}_{i,j}}$, $\begin{matrix} \tilde{\Omega}_{i,j} = \Omega_{i,j} - 2\varepsilon_{i,j,k}\omega_k \\ \Omega_{i,j} = \overline{\Omega}_{i,j} - \varepsilon_{i,j,k}\omega_k \end{matrix}$, $\overline{\Omega}_{i,j}$: time-averaged rotation rate in the reference frame of the angular speed ω_k	
source term of ε	$c_1 \rho S\varepsilon - c_2 \rho \frac{\varepsilon^2}{k + \sqrt{v\varepsilon}}$
σ_k	1.0
σ_ε	1.2
c_1	$\max\left\{0.43, \frac{\tilde{\eta}}{5+\tilde{\eta}}\right\}$
where $\tilde{\eta} = \frac{Sk}{\varepsilon}$, $S = \left(2S_{i,j}S_{i,j}\right)^{1/2}$, $S_{i,j} = \frac{1}{2}\left(\frac{\partial u_i}{\partial x_j} + \frac{\partial u_j}{\partial x_i}\right)$	

One of the weaknesses of the standard $k - \varepsilon$ model lies with the modeled equation for the dissipation rate. The well-known round-jet anomaly is considered to be mainly due to the modeled dissipation equation. The realizable $k - \varepsilon$ model was intended to modify this discrepancy. The differences between them are as given in Table 3.7. Numerical results are in good agreement with experimental results in cases where the flow features include strong streamline curvature, vortices, and rotation.

3.3.2 Combustion Model

Gas-phase turbulent combustion involves two time scales: chemical kinetic time scale and turbulent mixing time scale. The former means time requirement for complete reaction of certain gas species, while the latter means time requirement for breakup of large-scale eddies. The effect of turbulent mixing can be neglected when reaction components are premixed or turbulent mixing time is much shorter than reaction time. Then, the finite-rate reaction method can be used with relevant reaction rates based on averaged flow parameters. The effects of turbulent mixing and chemical kinetics must be simultaneously under consideration when their time scales are equivalent. Then combined the eddy-breakup model proposed by Spalding with finite-rate reaction method can be used. The effect of chemical kinetics can be neglected when turbulent mixing time is much longer than reaction time. Mixing fraction method can be used. Details of various simulation methods of turbulent combustion are in Ref. [45].

The flame of gas combustion in pulverized coal furnace, which is often simulated with non-premixed combustion model, is turbulently diffusive. Non-premixed combustion model involves the solution of transport equations for one or two conserved scalars (the mixture fractions) instead of all species. Then species concentrations are derived from the predicted mixture fraction fields. Interaction of turbulence and chemistry is accounted for with an assumed-shape probability density function (PDF). The non-premixed model has been specifically developed for the simulation of turbulent diffusion flames with fast chemistry. For such systems, the method is better than finite-rate reaction method combined with eddy-breakup model. It allows intermediate (radical) species prediction, dissociation effects, and rigorous turbulence-chemistry coupling, and is computationally efficient because it does not require the solution of a large number of species transport equations. Details of this method are in Ref. [46].

3.3.3 Radiative Heat Transfer Model

The main heat transfer way in pulverized coal furnaces is heat radiation. Radiative heat transfer between gas and particles must be considered for the existence of pulverized coal particles. P-1 radiation model and DO (discrete ordinates) radiation model can be adopted. P-1 radiation model assumes that all surfaces are diffuse and radiations are gray. For the P-1 model, the radiative transfer equation is a diffusion equation, which is easy to solve with little CPU demand. The model includes the effect of scattering. For combustion applications where the optical thickness is large, the P-1 model works reasonably well. The DO model spans the entire range of optical thickness and can be applied to solve various problems such as non-gray radiation, radiation at specular or semitransparent walls, and so on. But highly precise results demand more computational cost and memory requirements. Details of various radiative heat transfer models are in Ref. [46].

3.3.4 Discrete Phase Model

The flow of gas and particles in the pulverized coal furnace is a kind of dilute multiphase flow. There exists certain mass and volume distribution of these particles which would go through devolatilization and combustion in motion. Three methods are commonly used in the simulation: Eulerian–Eulerian (E-E) method, Eulerian–

Lagrangian (E-L) method, and probabilistic formulation method. In E-E method, the gas phase and particle phase are both treated as spatial and temporal continuous phase. In E-L method, the gas phase is treated as a continuum by solving the time-averaged Navier–Stokes equations, while the dispersed phase is solved by tracking a large number of particles through the calculated flow field. The dispersed phase can exchange momentum, mass, and energy with the gas phase. A typical instance is the widely used PSIC (particle-source-in-cell) method, where particle stress, heat transfer, and weight loss of particles due to evaporation, devolatilization, and char oxidation are calculated on each point of the trajectories coupled with the gas-phase characters. That is to say, the momentum, mass, and energy exchange between gas phase and dispersed phase are treated as source terms for gas-phase discrete equations and dispersed-phase ordinary differential equations. In probabilistic formulation method, the distributions of particles in the continuous phase are calculated with PDF (probability density function). The motion pattern of pulverized coal in the furnace is often investigated by E-L method which is in favor of the simulation of pulverized coal combustion process and superior to E-E method.

3.3.5 Reaction Models of Particles

When pulverized coal enters the furnace, it will experience the stages of heat-up, evaporation, devolatilization, char oxidation, burning, and burnout. For the simulation of pulverized coal combustion process by E-L method, additional reaction models of particles, including particle devolatilization model and char combustion model, need to be introduced. There are a large number of coal devolatilization models such as the single kinetic rate model, the two-competing-rate (Kobayashi) model, the generalized devolatilization model, the FG-DVC (functional group-depolymerization, vaporization, and cross-linking) model, the FLASHCHAIN model, and the CPD (chemical percolation devolatilization) model in many articles. The last three models all are based on the coal structure to describe the formation of the precursor tar by the network statistic method of simplified coal chemistry except their differences in network geometry, bridge-breaking and cross-linking, pyrolysis products, mass transfer assumption, and statistical approach. Though the two-competing-rate model is more commonly used than others at present, more advanced and accurate devolatilization models based on the coal structure are more favorable as the accuracy of pyrolysis simulation is directly relevant to the subsequent simulation of gas-phase combustion, particle movement, and pollutant formation. The CPD model has been introduced into the comprehensive numerical simulation of combustion in the furnace as a submodel. The calculated results are also in comparison with the experimental results [45]. The oxidation of char is a kind of heterogeneous surface combustion, which can be simulated with the diffusion-limited rate model, the kinetics/diffusion-limited rate model, and the intrinsic model which is based on the measured intrinsic char kinetic rate and the cavity diffusion model.

3.3.6 Pollutant Formation Model

Mostly concerned problem of NO_x formation modeling is the concentration distribution of NO inside the domain. The concentration distribution of NO is calculated by solving the mass transfer equation of NO with the consideration of the effect of convection, diffusion, formation, and reduction on the NO concentration. Simplest approach is the so-called "postprocessor," where the calculation of NO formation is performed after that of the combustion issue uncoupled with the flow and combustion, due to the extremely low NO concentration inside the domain. The source term in the NO governing equation represents the formation and reduction of NO. Details of the mechanism of NO_x formation and reduction are in Ref. [45, 46].

The direct effect of the interaction of turbulence and chemistry on the NO_x formation is significant. As the chemical kinetic time scale and the turbulent mixing time scale are in the same order of magnitude in every homogeneous reaction of NO formation, the effect of turbulent fluctuation on the chemical reaction rate must be considered, which is often introduced into species conservation equations by the time-averaged method. But generally time-averaged reaction rate cannot be derived from time-averaged temperature and concentration for the high nonlinearity between them. There are many approaches proposed to take the effect of turbulent fluctuation on the reaction rate into account. Following approaches are in the order of complexity: (1) To directly calculate the reaction rate based on the parameters which are obtained by solving the time-averaged Navier–Stokes equations is the simplest but not quite accurate approach, because it neglects the effect of turbulence on the chemistry. However, this approach is the mostly used in the commercial-scale simulation of NO_x formation in the utility boiler furnace. (2) Moment closure method. (3) Monte Carlo PDF method: This method can describe complex reaction processes and accurately simulate the interaction between turbulence and chemistry. It is only suitable for simplified chemical mechanism due to its huge computational cost. (4) LES: The simulation of NO_x formation by LES has not yet been mature enough in the simulation of practical combustion system. (5) DNS: The simulation of NO_x formation by DNS in practical combustion simulation is not possible so far. It is often used in the theoretical research of fundamental turbulence database.

3.3.7 Application Examples

Likun Huang and Zhengqi Li from Harbin Institute of Technology in China numerically studied the effect of the overfire air on the air flow and coal combustion in a 670 t/h wall-fired boiler [47]. The comparison between the numerical results of the cold air flow and the experimental data of a 670 t/h coal-fired boiler shows a good coincidence. The NO value of calculation is 12% less than the ones of measurement but the CO value is almost same between calculation and measurement.

Fig. 3.57 Distribution of the averaged temperature and oxygen mass fraction at the furnace height [47]

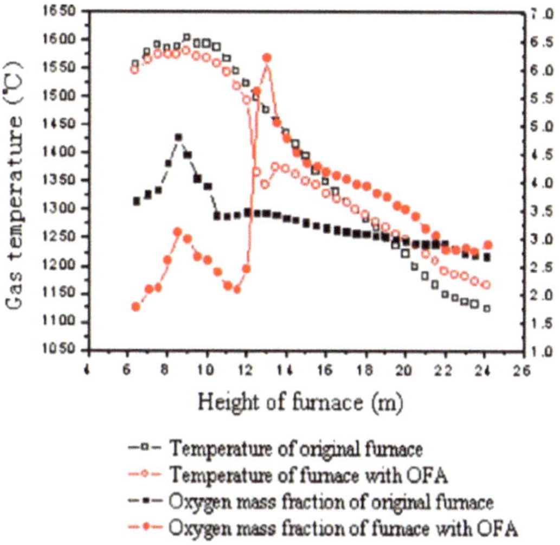

Fig. 3.58 Distribution of the averaged NO concentration at the furnace height [47]

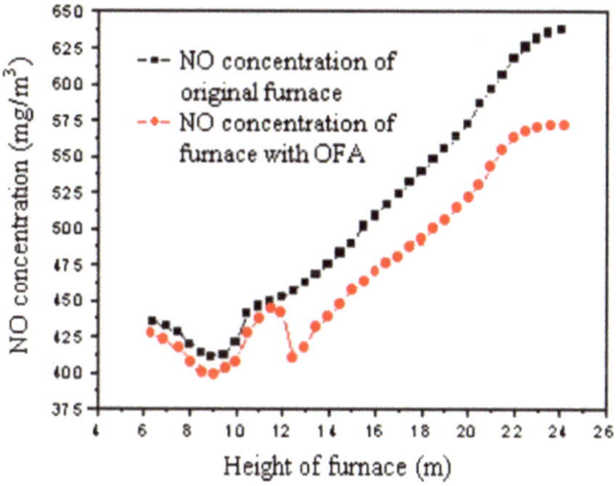

Mathematical models and calculation method are proved to be valid. The aerodynamic field in the furnace with various overfire air (OFA) parameters and arrangements is calculated to analyze the effect of OFA after the OFA ports are added to the boiler furnace. The optimized OFA parameters and arrangements are gained and simulated in reacting flow. The result shows that the NO emission of the furnace with OFA is reduced, and the NO concentration of outlet decreases from 638 to 570 mg/Nm^3 ($O_2 = 6\%$) (Figs. 3.57, 3.58, and 3.59).

Vikhansky and Bar-Ziv from Ben-Gurion University of the Negev in Israel use a 3-D CFD code to analyze the performance of 550 MW pulverized coal combustion opposite a wall-fired boiler (of IEC) at different operation modes [48]. The main objective of this study was to prove that connecting plant measurements with 3-D furnace modeling is a cost-effective method for design, optimization, and

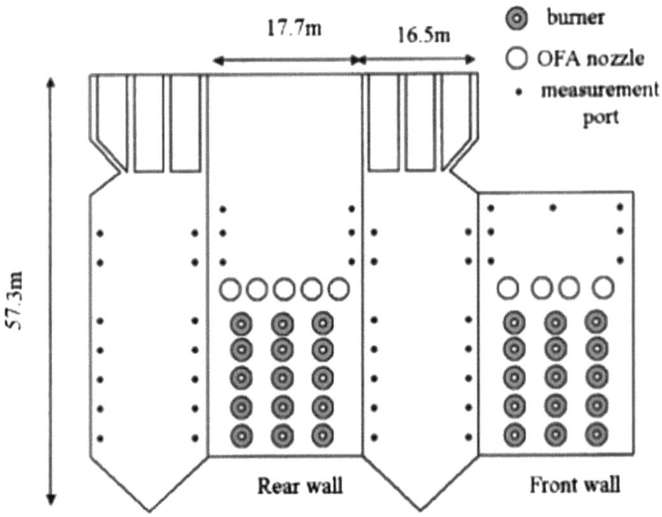

Fig. 3.59 Schematic view of the boiler and its sizes. The boiler is equipped with a set of heat flux probes that measure incident radiation heat fluxes. The probes are installed in the measurement ports and at the burners [48]

problem-solving in power plant operation. The computational model includes a 3-D turbulent reacting flow, radiative heat transfer, and Lagrangian particle tracking. The interactions between gas phase, radiation fields, and particles are accounted for. Turbulence is modeled by Favre-averaged $k - \varepsilon$ model. Gas-phase reactions are calculated assuming local instantaneous equilibrium. The effect of turbulence on chemical composition is incorporated by the assumption that the mixture fraction is described by a prescribed probability density function. Particles are tracked by calculating the mean trajectory and cloud diameter for each discrete group of particles in a Lagrangian frame of reference. Heat flux results from calculations were compared with measurements from the boiler, which showed a good agreement. Consequently, the code was used to study hydrodynamic aspects of air–flue gases mixing in the upper part of the boiler. It was demonstrated that effective mixing between flue gases and overfire air is of essential importance for CO reburning (Figs. 3.60, 3.61, and 3.62).

Lingyan Zeng and Zhengqi Li from Harbin Institute of Technology in China experimentally and numerically studied the pulverized coal combustion process and NO_x emissions for a 300 MWe wall-fired boiler burning bituminous coal [49]. Measurements of the polar components of the gas velocity made with a particle-dynamics anemometer and calculated values are found to be in good agreement. Turbulence was modeled by the realizable $k - \varepsilon$ model. The Lagrangian stochastic tracking model was applied to analyze the gas/particles flow field, while calculations of gas/particle two-phase coupling employed the particle-source-in-cell (PSIC) method. Radiation was modeled using the P-1 model. Devolatilization was modeled with the two-competing-rate model. The combustion of volatiles was

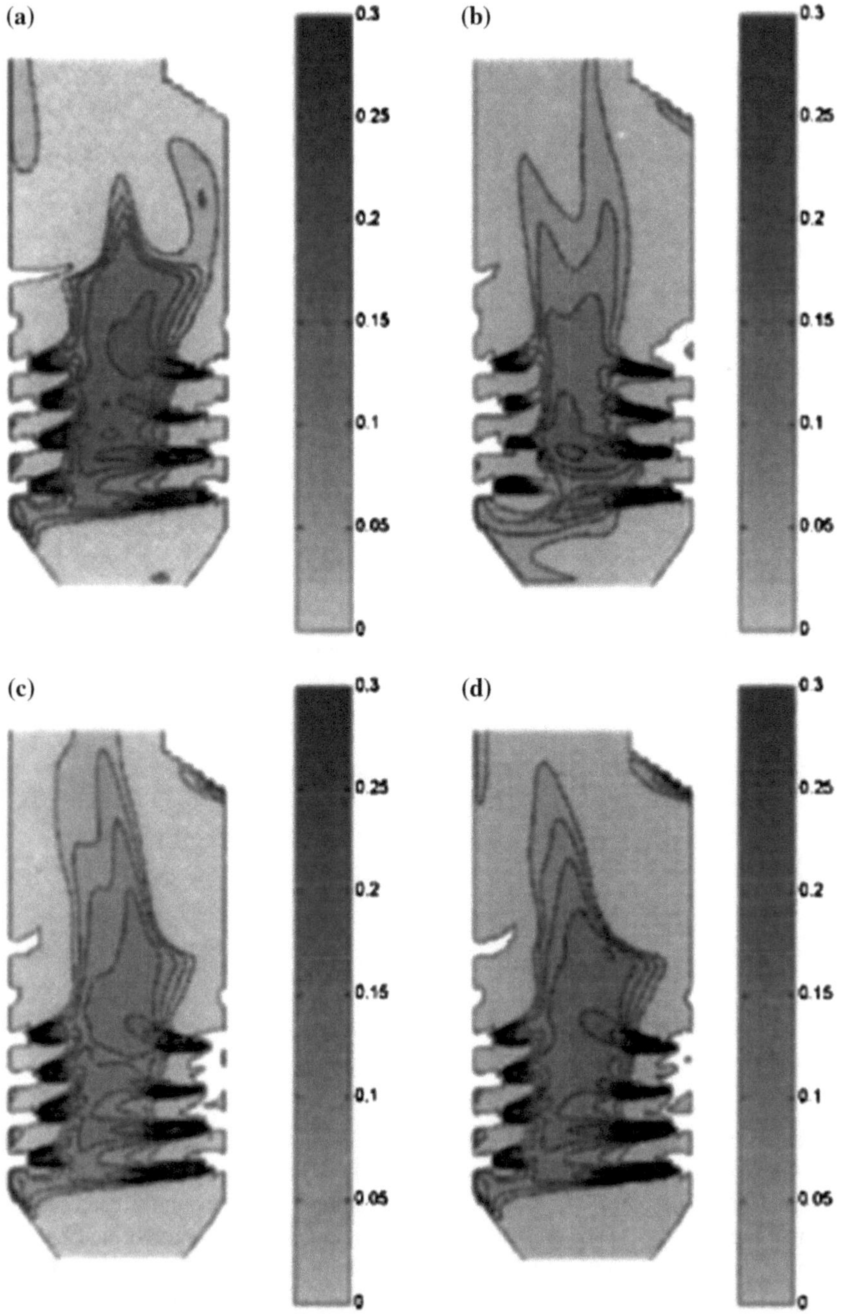

Fig. 3.60 Distribution of molar fraction of CO at the symmetric plane of the boiler: **a** test "A"; **b** test "B"; **c** case "A1"; and **d** case "A2." As one can see CO concentration in the lower part of the furnace is higher in case "A," but it undergoes abrupt change at the OFA level. In case "B", CO concentration is lower but it has a long tail that goes out of the furnace zone [48]

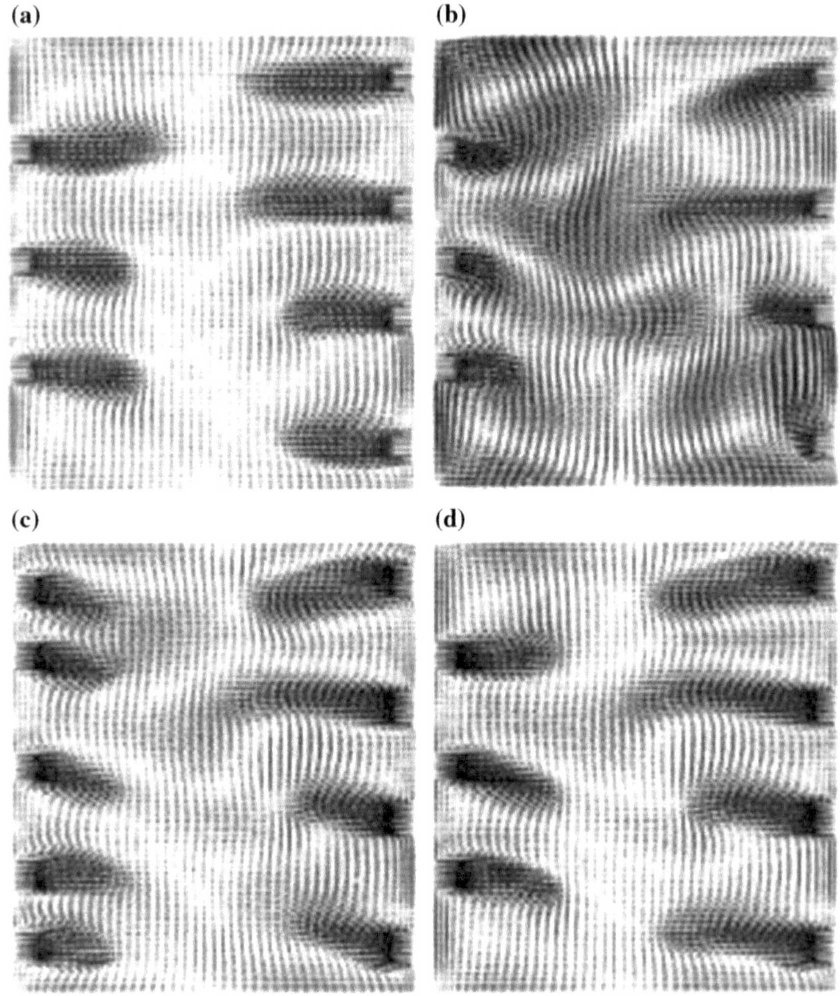

Fig. 3.61 Velocity vectors plots at the OFA nozzles level for test "A" (**a**), test "B" (**b**), test "A1" (**c**), and test "A 2" (**d**) [48]

modeled by employing a probability density function theory and char combustion by employing the diffusion/kinetics model. The formation of NO_x only includes thermal NO and fuel NO. As the outer secondary air vane angle increases, the temperature of the water-cooled tube wall region near the burners and furnace hopper increases. Changes in the NO_x concentration at the furnace outlet are not simply linear with changes in the outer secondary air vane angle. The NO_x concentration at the furnace outlet was lowest (420.2 ppm at 6% O_2) for an outer secondary air vane angle of 35° and highest (468.2 ppm at 6% O_2) for an outer

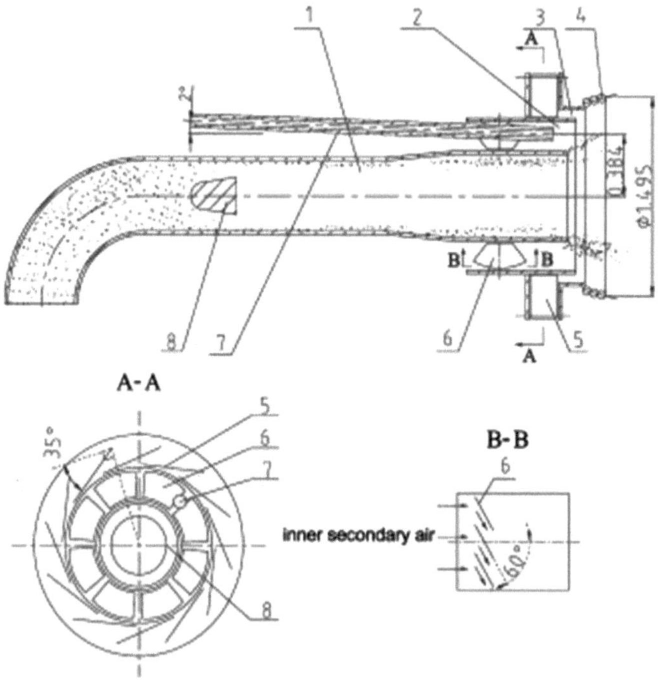

Fig. 3.62 EI-DR burner and the position of the monitoring pipe (dimensions in meters): (1) primary air duct, (2) inner secondary air duct, (3) outer secondary air duct, (4) water-cooled wall, (5) tangential vanes, (6) radial vanes, (7) monitoring pipe, and (8) conical diffuser [49]

secondary air vane angle of 30°. This represents an increase of 11.4% (Figs. 3.63, 3.64, 3.65, 3.66, and 3.67).

Hari Badu Vuthaluru and Rupa Vuthaluru from Curtin University of Technology in Australia develop a 3-D combustor model for predicting the performance of full-scale tangential fired (TF) boiler and determine the flow patterns of the gas and coal particles, with an emphasis on formation mechanism of gas flow deviations and uneven temperature on the superheaters, reheaters, and divisional superheater sections of the furnace [50]. The importance of these simulations is to identify the locations of optimum additive injection ports to achieve maximum impact of additives in the combustion process to minimize the temperature deviation and reduce ash-related issues. This study is a classic example of numerical investigation into the problem of turbulent reacting flows in large-scale furnaces employed in thermal power plants for the remediation of ash deposition problems. The gas flow is described by the time-averaged equations of global mass, momentum, enthalpy, and species mass fraction. The particle phase equations were formulated in Lagrangian form, and the coupling between phases is introduced through particle sources of Eulerian gas-phase equations. The standard $k-\varepsilon$ turbulence model, single mixture fraction probability density function (PDF), and the P-1 radiation

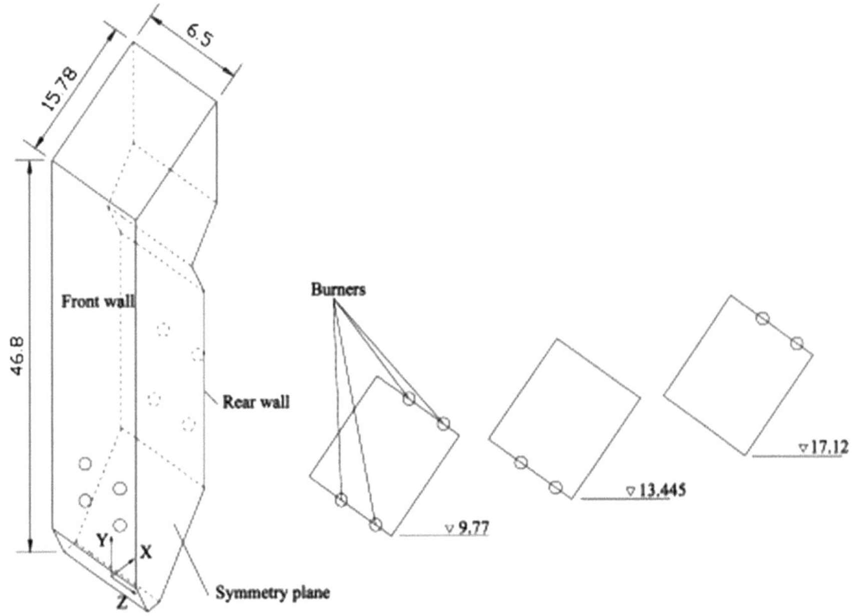

Fig. 3.63 Schematic diagrams of the structure of half of the furnace (dimensions in meters) [49]

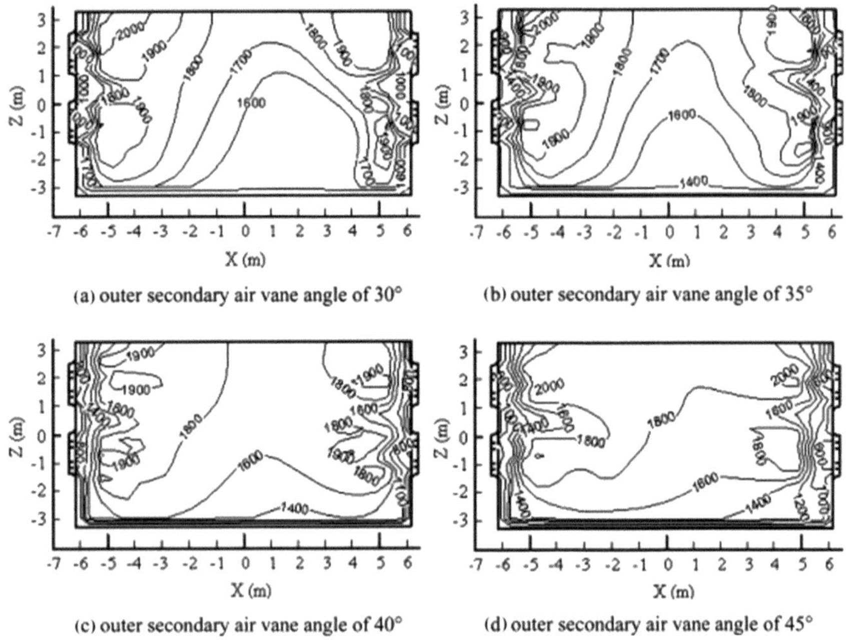

Fig. 3.64 Calculated temperature fields over a cross section through the burner center at a height of 9.77 m (dimensions in Kelvin) [49]

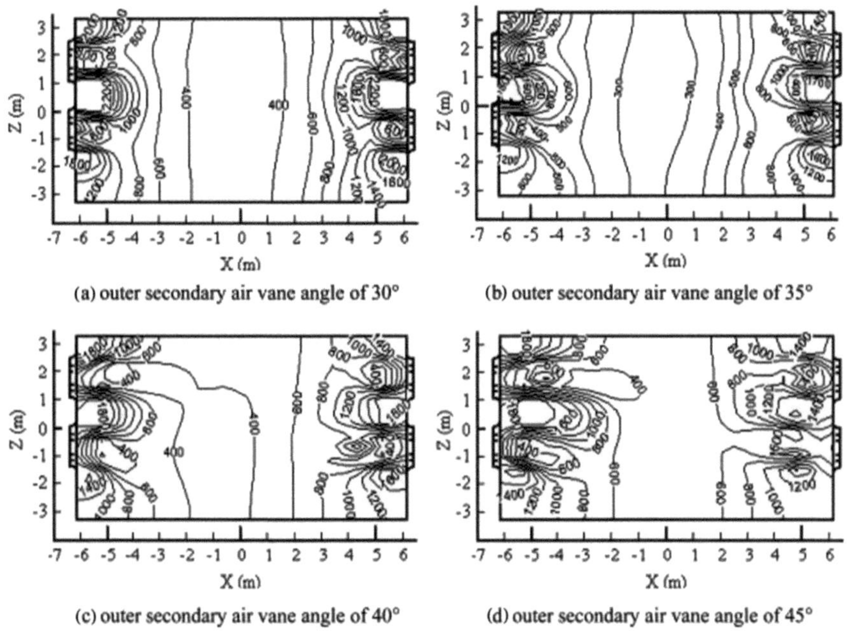

Fig. 3.65 Calculated NO_x concentration (ppm) over a cross section through the burner center at a height of 9.77 m [49]

models are used in the simulations. Excellent agreement between the simulation results and key boiler design values and available site operation records following full-scale trials indicate that the calculations are reliable (Figs. 3.68, 3.69, and 3.70).

Luis I. Díez and Cristóbal Cortés from University of Zaragoza in Spain carried out a CFD investigation about the performance of a 600 MWe tangentially coal-fired boiler, focusing on the reduction of NO_x attainable by using overfire air [51]. A comprehensive combination of NO_x chemistry models has been used, coupled with the numerical simulation of fluid and particle flow, solid fuel combustion, and heat and mass transfer. Turbulence was modeled by the standard $k - \varepsilon$ model. Devolatilization was modeled with the single kinetic rate model and the two-competing-rate model. The gas-phase homogeneous combustion was modeled by employing the non-premixed combustion model with a PDF theory, and char heterogeneous combustion was modeled by employing the double-film model. The Lagrangian stochastic tracking model was applied to analyze the gas/particles flow field. Radiation was described using the discrete transfer model. The formation of NO_x only includes thermal NO and fuel NO. Predicted values of gas temperature and species concentration have been adopted to validate the model against actual measurements from the full-scale boiler, under conventional and overfire air arrangements. A reasonable agreement has been attained in most cases (Figs. 3.71, 3.72, and 3.73).

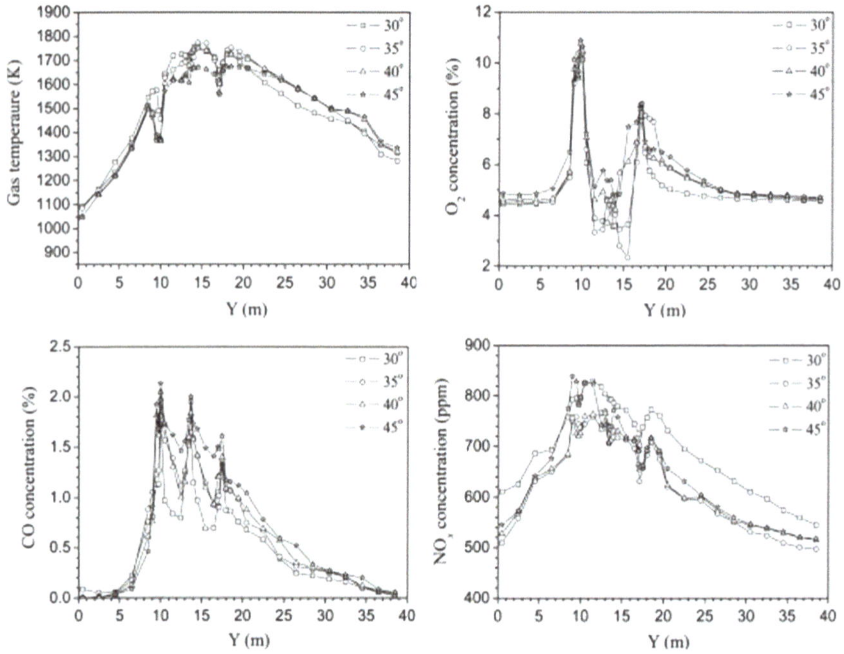

Fig. 3.66 Distributions of the average gas temperatures and concentrations along the furnace height for the four vane angles [49]

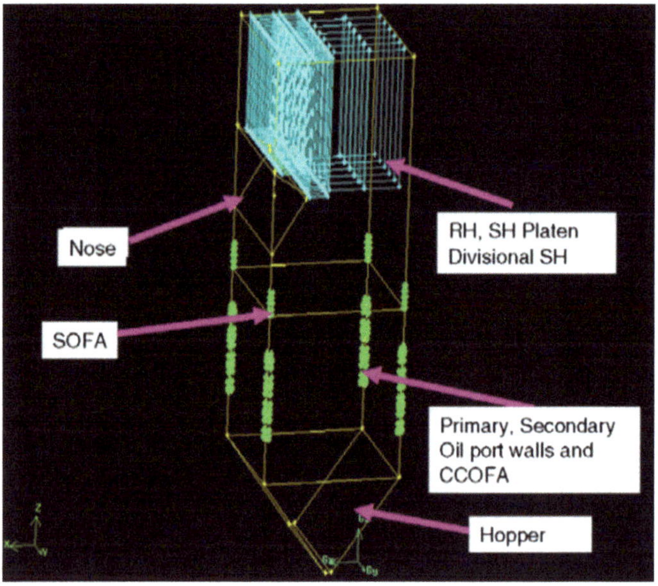

Fig. 3.67 Furnace geometry [50]

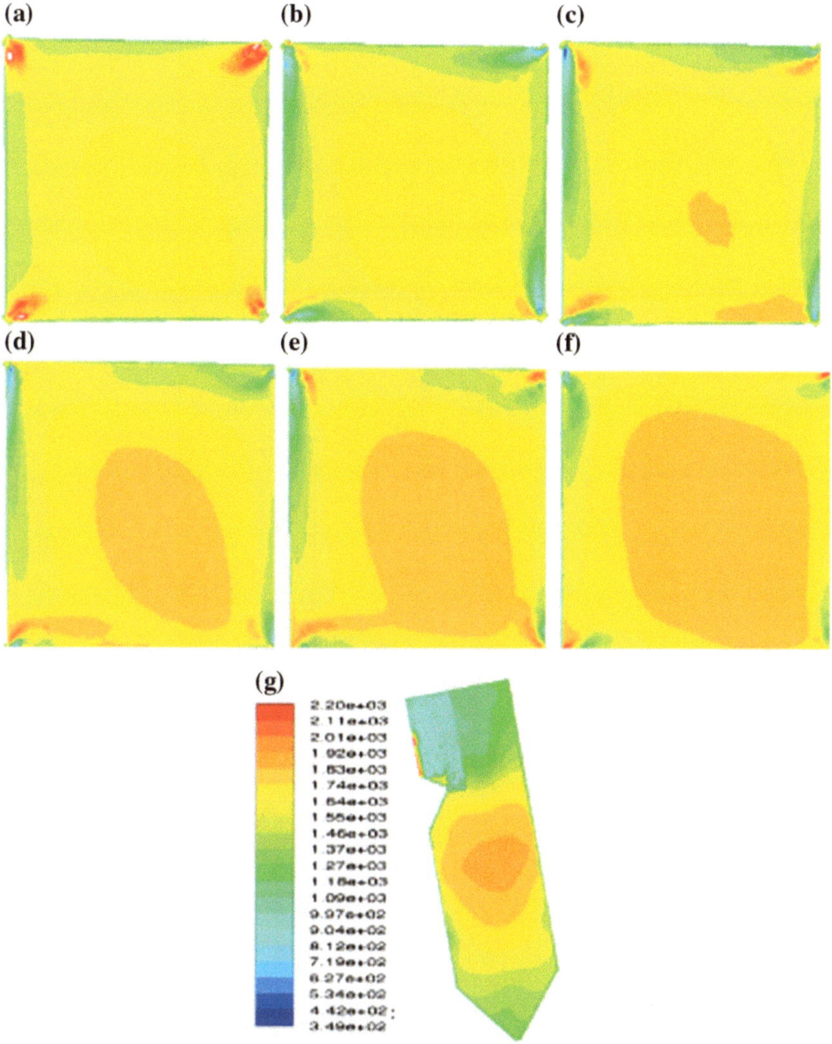

Fig. 3.68 Contours of temperature for case 5 (50% Kideco–50% Berau with all burners switched on). **a** Burner A; **b** burner B; **c** burner C; **d** burner D; **e** burner E; **f** burner F; and **g** $Y = 10$ m plane [50]

Changqing Dong and Yongpin Yang from North China Electric Power University in China carried out CFD modeling of product gas (from biomass gasification) and coal cofiring in a 600 MWe tangential PC boiler [52]. Compared to directly cofiring of biomass and coal, it might avoid feeding biomass into boiler, reduce boiler fouling and corrosion problem, and avoid altering ash characteristics.

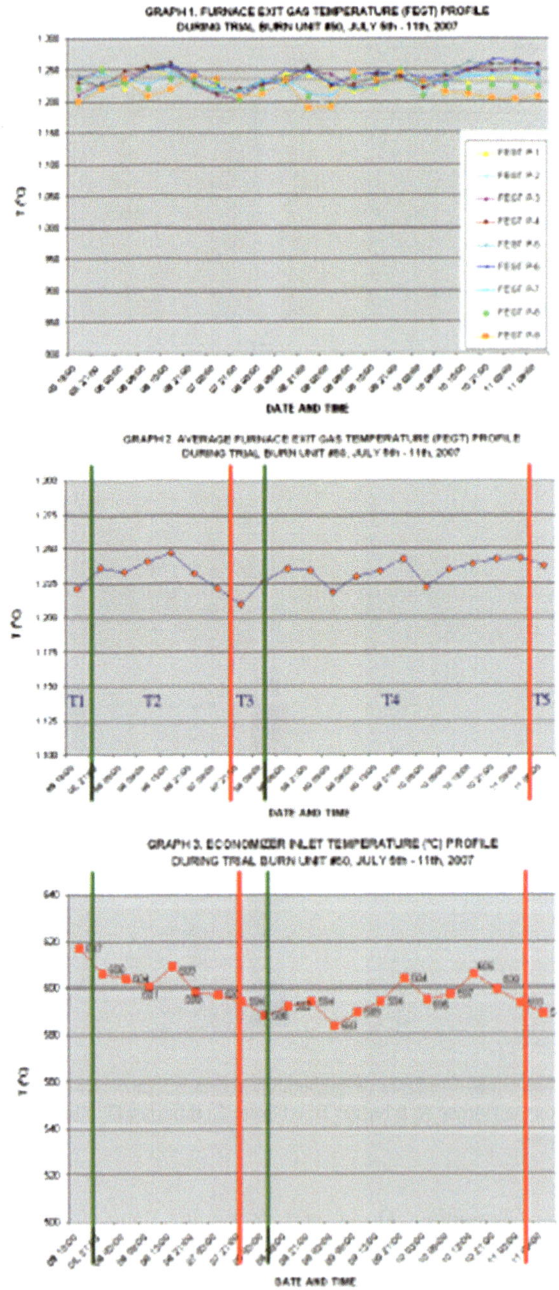

Fig. 3.69 Observations from trial runs with bar lines showing the trial schedules T1, T2, T3, T4, T5, and T5 [50]

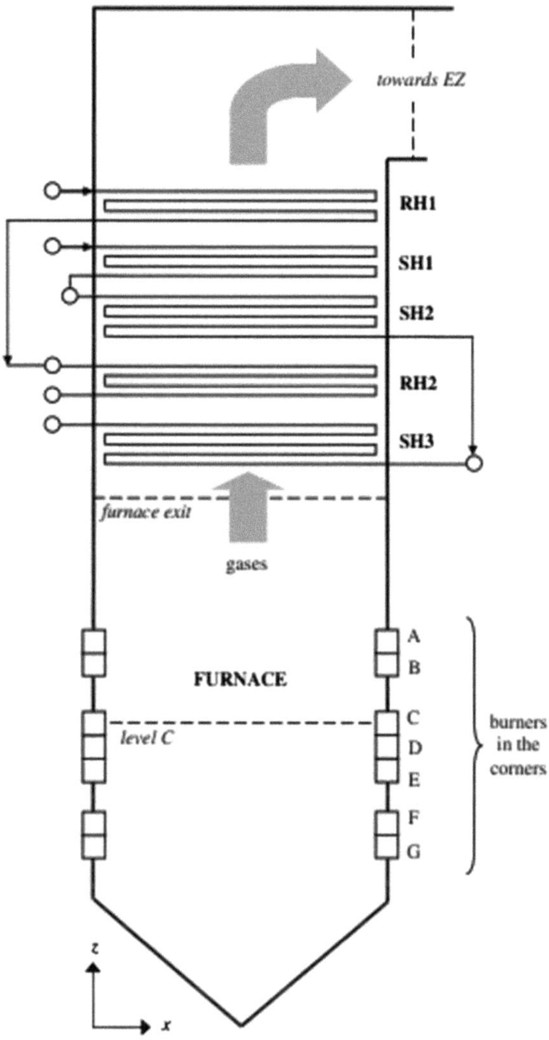

Fig. 3.70 Schematic arrangement of burners (levels A–G) and heat recovery sections in the case-study boiler [51]

Turbulence was modeled by the standard $k - \varepsilon$ model. The dispersion of particles due to turbulence in the fluid phase was predicted using the stochastic tracking model. Radiation was modeled using the P-1 model. The two-competing-rate model was selected as the devolatilization model, two-mixture fraction/PDF as the gas turbulent combustion model, and the kinetic/diffusion surface reaction rate model as the surface combustion model. The formation and reduction of NO was considered as fast reaction. NO emission was predicted by thermo-chemical calculation in chemical equilibrium model. The results showed that NO_x emission was reduced

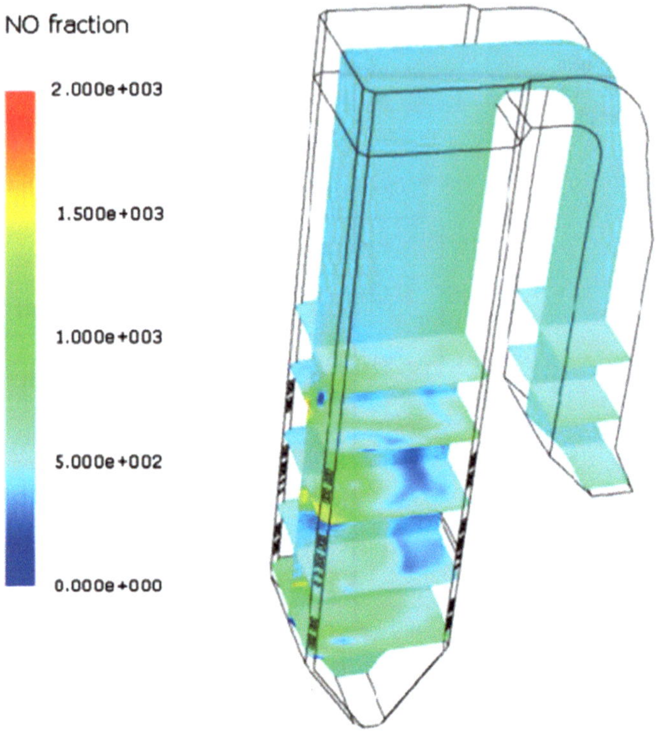

Fig. 3.71 Prediction of NO*x* distribution throughout the computational domain, for the conventional operating scenario [51]

about 50–70% when the product gas was injected through the lowest layer burner. The fouling problem can be reduced with furnace temperature decreasing for cofiring case. The convection heat transfer area should be increased or the cofiring ratio of product gas should be decreased to keep boiler rated capacity (Figs. 3.74, 3.75, 3.76, and 3.77).

Norbert Modlinski from Wroclaw University of Technology in Poland carried out a 3-D numerical investigation of OP-380 boiler tangentially fired furnace utilizing bituminous, which was retrofitted by replacing traditional jet burners with RI-JET2 (Rapid Ignition-JET) swirl burners [53]. Turbulence was modeled by the realizable $k - \varepsilon$ model. Discrete phase trajectories were calculated using Lagrangian formulation, and the coupling between the phases was introduced through particle sources of Eulerian gas-phase equations. A discrete ordinates method was used to evaluate radiative heat transfer. Devolatilization was modeled with the single kinetic rate model which was extended by using FG-DVC network devolatilization model as a preprocessor. Gas-phase combustion was modeled using the species

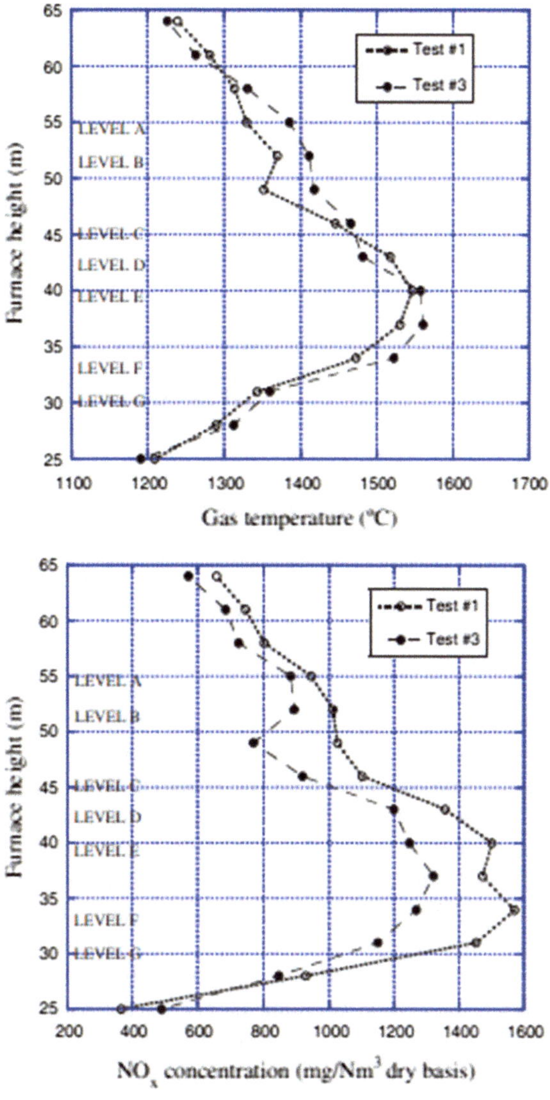

Fig. 3.72 Predicted cross-sectional averages of gas temperature and NO_x concentration along the furnace height [51]

transport model with a four-step global mechanism, the turbulence-chemistry interaction using the eddy-dissipation concept (EDC), and char combustion using the diffusion-limited rate model. An inertial impaction mechanism of fly ash deposition was considered only to predict the water walls slagging process. Results were compared to simulations of similarly designed boiler equipped with traditional jet burners. Stable flame operation was noticed even without secondary and tertiary air swirl. 3-D simulations of combustion chamber have shown that in a burner zone, a visibly isolated, concentrated flame exists in the furnace axis. This kind of flame

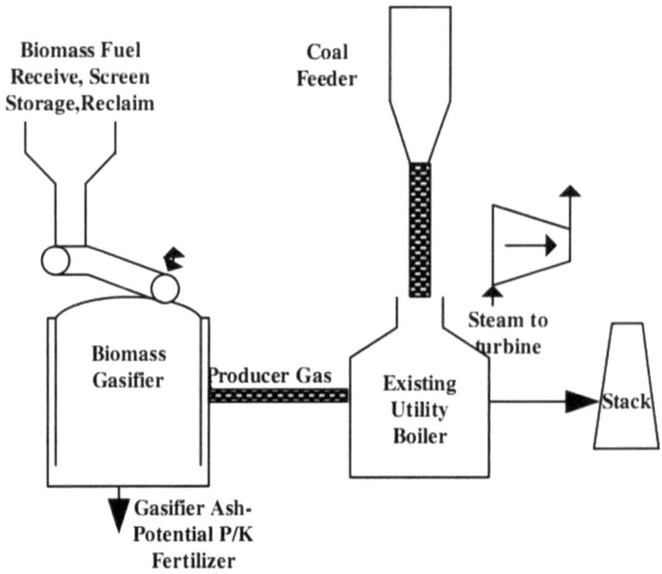

Fig. 3.73 Biomass cofiring system [52]

shape reduces corrosion risk and furnace walls slagging as a result of RI-JET2 burner's long range (Figs. 3.78, 3.79, 3.80, and 3.81).

Srdjan Belosevic and Miroslav Sijercic from Laboratory for Thermal Engineering and Energy in Serbia and Montenegro numerically simulated processes in utility boiler pulverized coal tangentially fired dry bottom furnace [54]. Five operation regimes of the 210 MWe boiler furnace burning Serbian lignite, with different grinding fitness of coal and coal quality, have been simulated. Turbulence was modeled by the standard $k-\varepsilon$ model. Dispersed phase was described by differential equations of motion, energy, and mass change in Lagrangian field, with diffusion mode of particle dispersion by turbulence, while calculations of gas/particle two-phase coupling employed the particle-source-in-cell (PSIC) method. The six-flux method was used to model thermal radiation. Char combustion was modeled by employing a diffusion/kinetics model. The model successfully predicts the influence of the parameters on the furnace processes and operation characteristics (like the flue gas temperature and the furnace walls radiation fluxes). The predicted flame temperature and percentage combustibles in bottom ash are in good agreement with the measurements (Figs. 3.82, 3.83, 3.84, 3.85, 3.86, and 3.87).

T. Asotani and T. Yamashita from Nagoya University in Japan studied the pulverized coal ignition behavior in a 40 MW tangentially fired commercial boiler [55]. The standard $k-\varepsilon$ model and Lagrangian particle tracking method with the random walk model were used for a gas-particle turbulence flow simulation. Radiation was modeled using the P-1 model. Devolatilization was modeled with the single kinetic rate model. Gas-phase combustion was modeled using the species

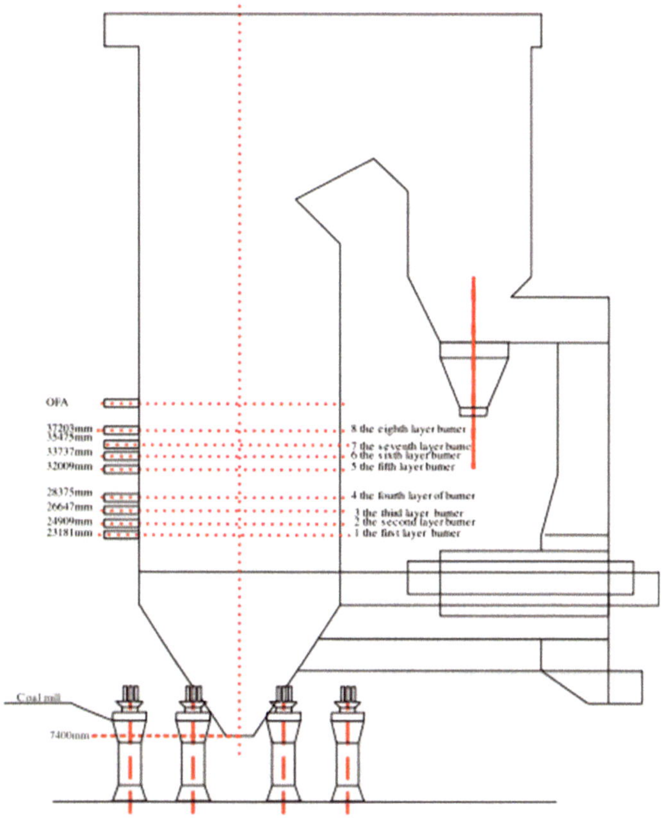

Fig. 3.74 The Yuanbaoshan boiler elevation [52]

Fig. 3.75 Mean temperature distribution along with the height of the furnace [52]

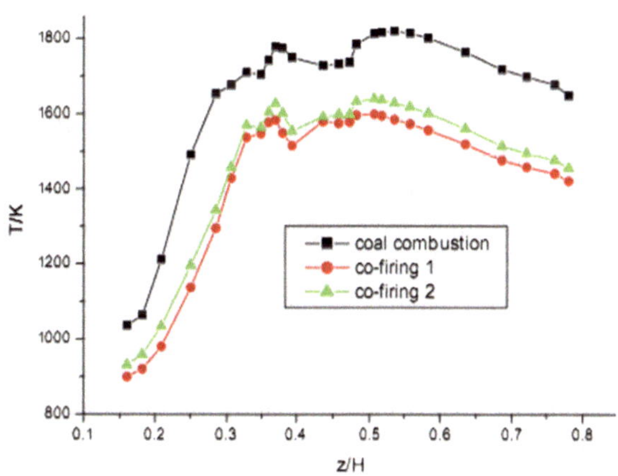

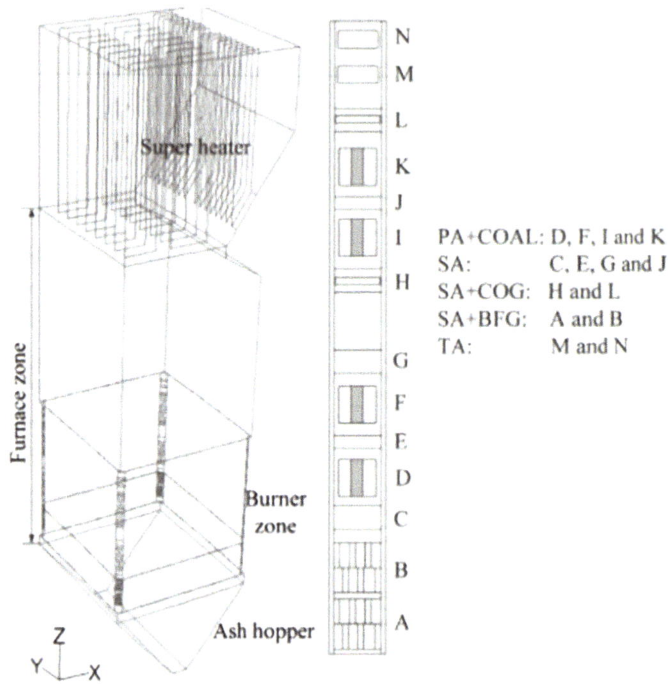

Fig. 3.92 Schematic diagram of the furnace and burner nozzle arrangement at each corner [57]

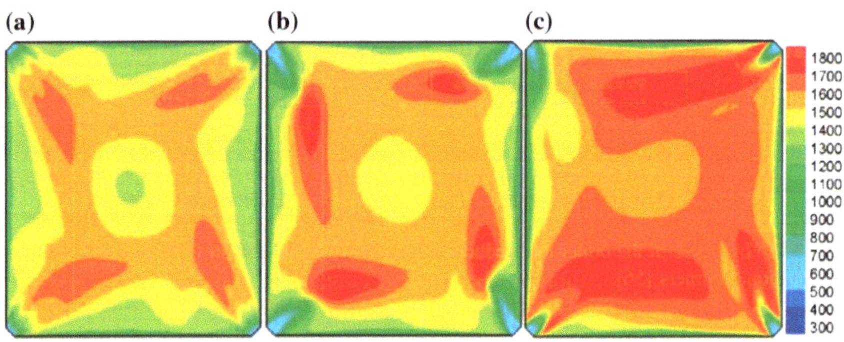

Fig. 3.93 Temperature distributions of central cross sections of **a** BFG, **b** pulverized coal, and **c** COG nozzles in case 1 (unit: K) [57]

flue gas added to the fuel gas are proposed and tested for predried coal with 55, 45, 35, 25, and 15% moisture contents. The increase in the furnace temperature for the recirculated flue gas arrangement is smaller than that of the additional air arrangement, and stack loss is expected to be smaller than for additional air cases. The temperature distributions and wall heat fluxes in the boiler for the recirculated

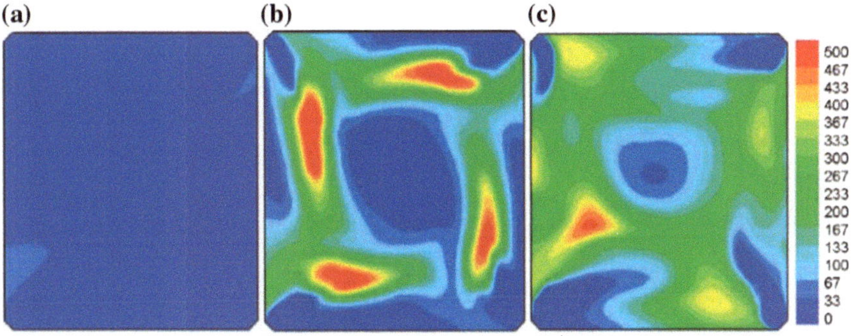

Fig. 3.94 NO concentration distributions of central cross sections of **a** BFG, **b** pulverized coal, and **c** COG nozzles in case 1 (unit: ppm) [57]

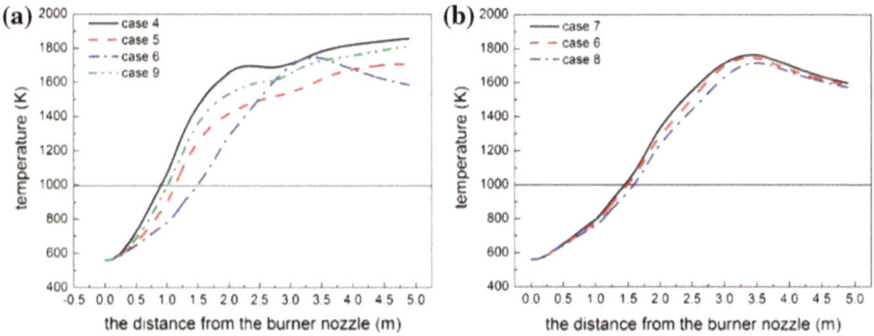

Fig. 3.95 Comparisons of variation of the temperature along the central line of the PA nozzle: **a** changing the BFG flow rate and **b** changing the COG flow rate [57]

flue gas cases are similar to that of the raw coal combustion case. Therefore, use of recirculated flue gas is proposed as an option for future operation of the furnace with predried coal (Figs. 3.102, 3.103, and 3.104).

Srdjan Belosevic and Vladimir Beljanski from Laboratory for Thermal Engineering and Energy in Serbia use CFD method to examine the efficiency of combustion modifications in the furnace of Kostolac-B 350 MWe boiler units, tangentially fired by pulverized lignite [60]. Selected predictions of the emission, the furnace exit gas temperature, NO concentration, gas temperature, and velocity field are given for the case-study furnace under different operating conditions. The individual or combined effects of coal and preheated air distribution over the individual burners and the burner tiers, the grinding fineness and quality of coal, and the cold air ingress were investigated. Turbulence was modeled by the standard $k - \varepsilon$ model. Dispersed phase was described by differential equations of motion, energy, and mass change in Lagrangian field, with diffusion mode of particle dispersion by turbulence, while calculations of gas/particle two-phase coupling employed the particle-source-in-cell (PSIC) method. The six-flux method was used

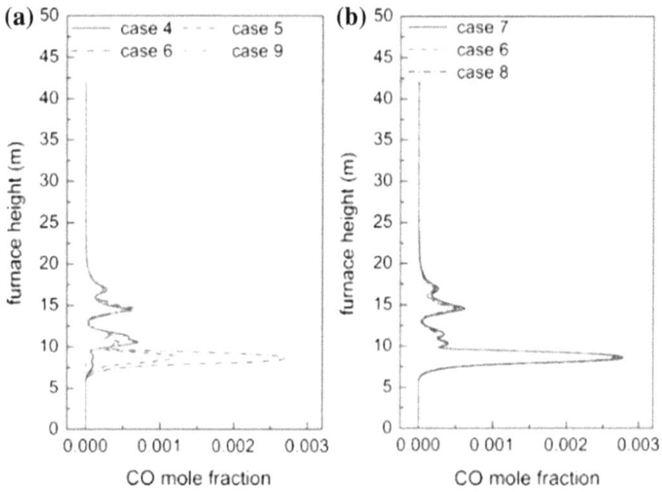

Fig. 3.96 Comparison of the average CO concentration profiles along the furnace height: **a** changing the BFG flow rate and **b** changing the COG flow rate [57]

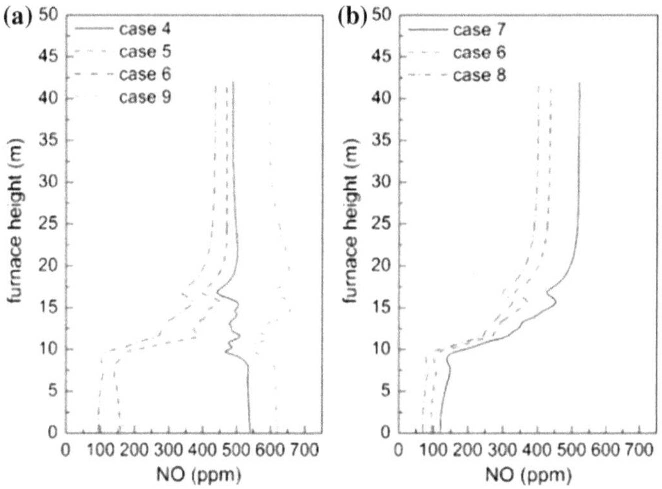

Fig. 3.97 Comparison of the average NO concentration profiles along the furnace height: **a** changing the BFG flow rate and **b** changing the COG flow rate [57]

to model thermal radiation. Char combustion was modeled by employing a diffusion/kinetics model. The formation of NO_x only includes thermal NO and fuel NO. Reduced emissions of up to 20–30% can be achieved only by proper organization of the combustion process. Obtained results were verified by the boiler thermal calculations. An optimal range of the furnace exit gas temperatures, 990 °C (995 °C)–1010 °C, was proposed, with respect to the safe operation of the steam

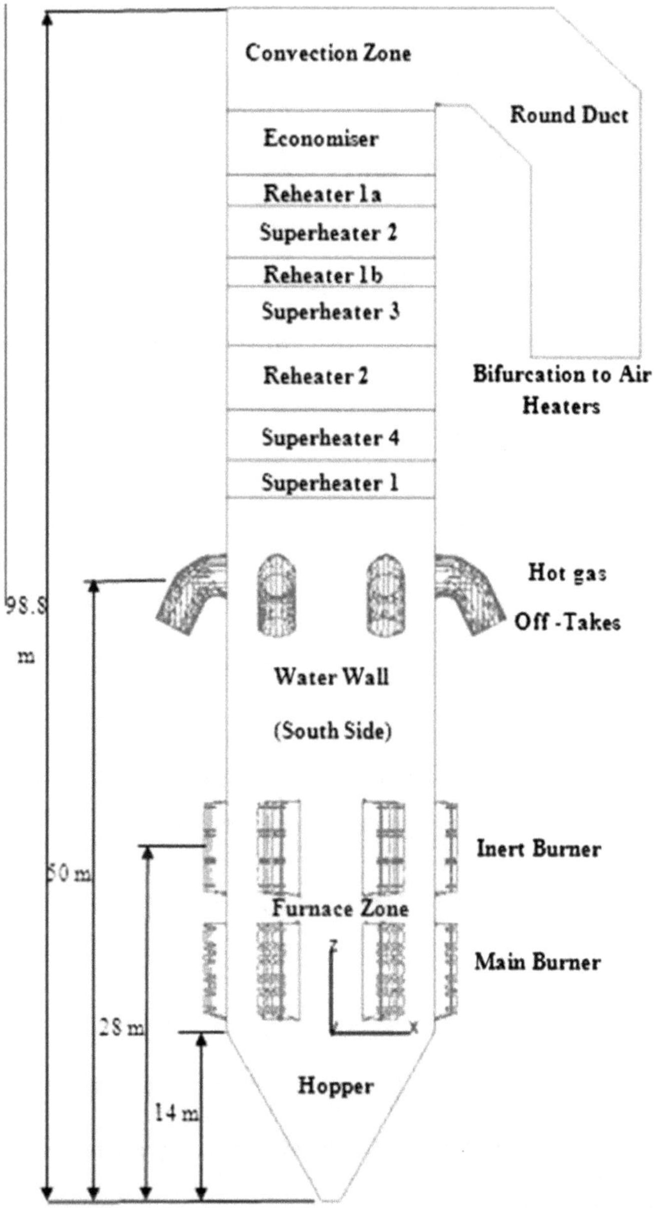

Fig. 3.98 The geometric description of the CFD model for the boiler, unit 1 at Loy Yang A power station [58]

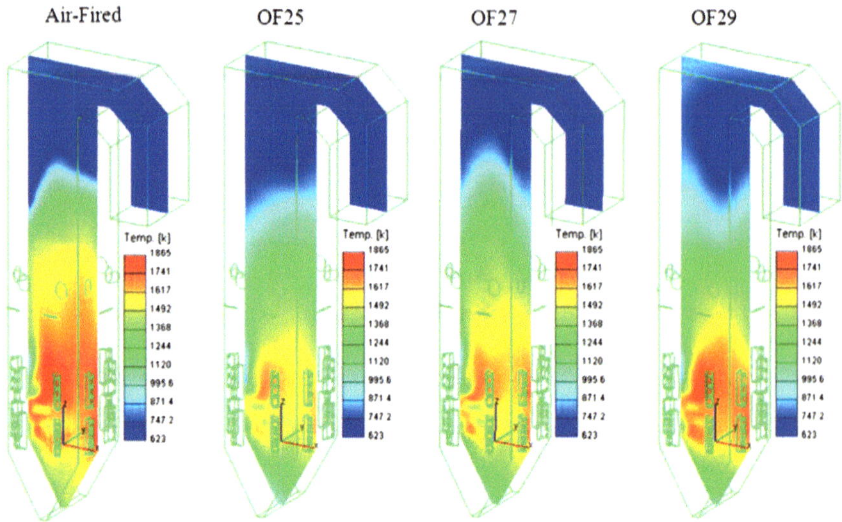

Fig. 3.99 Distributions of the flue gas temperature (K) along the height of the furnace at the midcut (*X–Z* plane) for air-fired, OF25, OF27, and OF29 combustion cases [58]

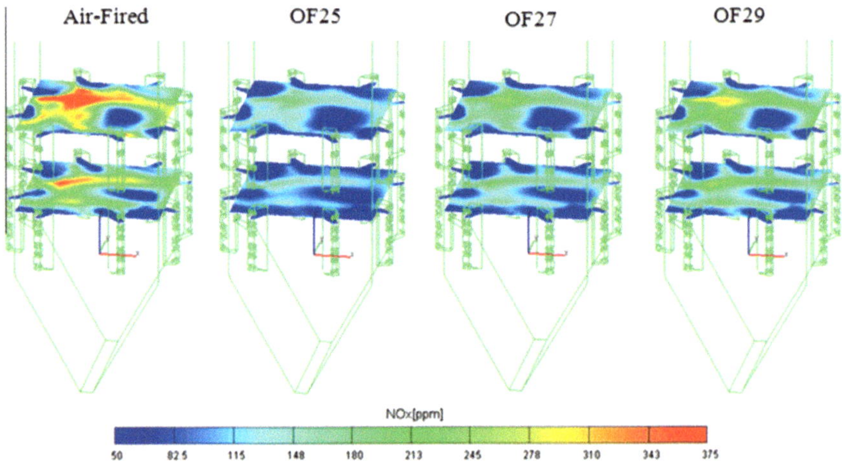

Fig. 3.100 Distributions of NO_x (ppm) at the UMB plane (lower *X–Y* plane in the figure) and at the UIB plane for all cases investigated [58]

superheater. Simulations by means of a computer code developed for the purpose showed that the air staging using overfire air ports might provide the NO_x emission reduction of up to 24% in the test cases with relatively high emission and up to 7% of additional reduction in already optimized cases (Figs. 3.105, 3.106, and 3.107).

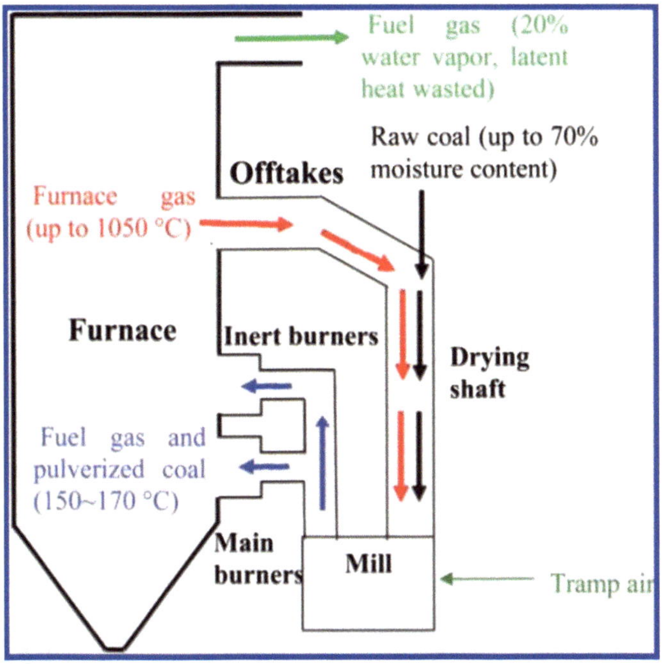

Fig. 3.101 Heating and drying processes of raw brown coal in the tangentially fired boilers in Latrobe Valley [59]

Choeng Ryul Choi and Chang Nyung Kim from Kyung Hee University in Republic of Korea numerically studied the characteristics of the flow, combustion, temperature, and NO_x emissions in a 500 MWe tangentially fired pulverized coal boiler with emphasis on fuel and thermal NO_x formations [61]. Turbulence was modeled by the RNG $k - \varepsilon$ model. The Lagrangian stochastic tracking model was applied to analyze the gas/particles flow field with their coupling. Radiation was modeled using the discrete ordinates radiation model. Devolatilization was modeled with the two-competing-rate model. The combustion of volatiles was modeled by employing the non-premixed combustion model with a PDF theory and char combustion by employing the diffusion/kinetics model. The formation of NO_x only includes thermal NO and fuel NO. The comparison between the measured values and predicted results shows good agreement, which implies that the adopted combustion and NO_x formation models are suitable for correctly predicting characteristics of the boiler. The relations among the predicted temperature, O_2, and CO_2 mass fractions are discussed based on the calculated distributions. The predicted results clearly show that NO_x formation within the boiler highly depends on the combustion processes as well as the temperature and species concentrations. The results obtained from this study have shown that overfire air (OFA) operation is an efficient way to reduce the NO_x emissions of the pulverized coal-fired boiler. Air staging combustion technology (OFA operation) adopted in this boiler has helped

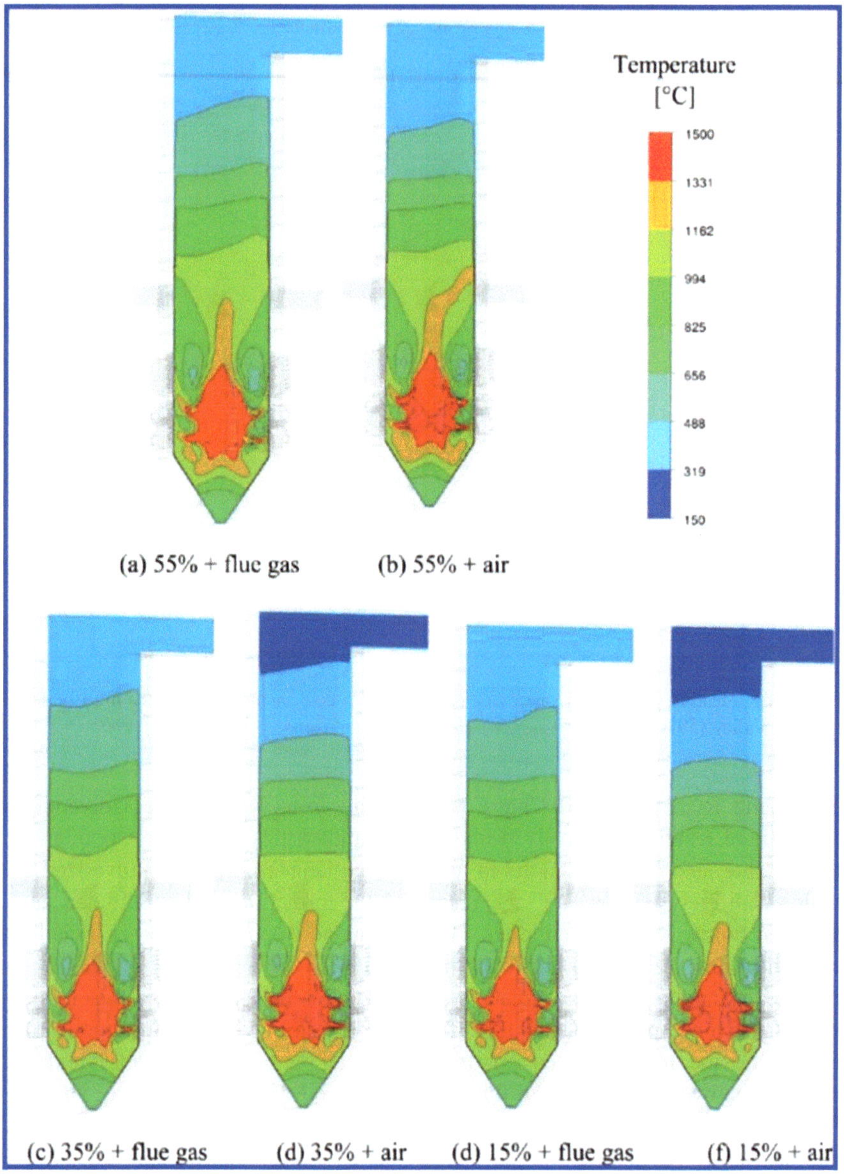

Fig. 3.102 Predicted gas temperature at the vertical midplane: **a** case 9, **b** case 4, **c** case 11, **d** case 6, **e** case 13, and **f** case 8 [59]

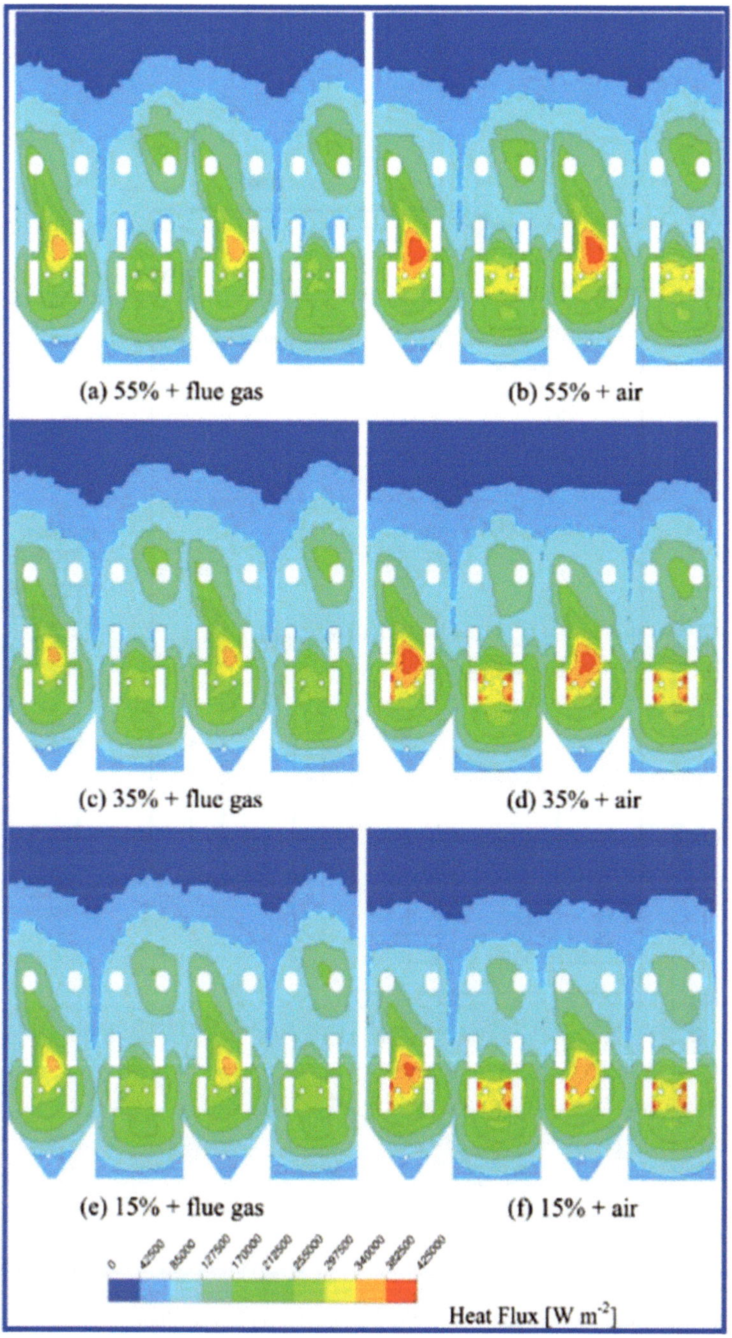

Fig. 3.103 Predicted wall incident heat flux: **a** case 9, **b** case 4, **c** case 11, **d** case 6, **e** case 13, and **f** case 8 [59]

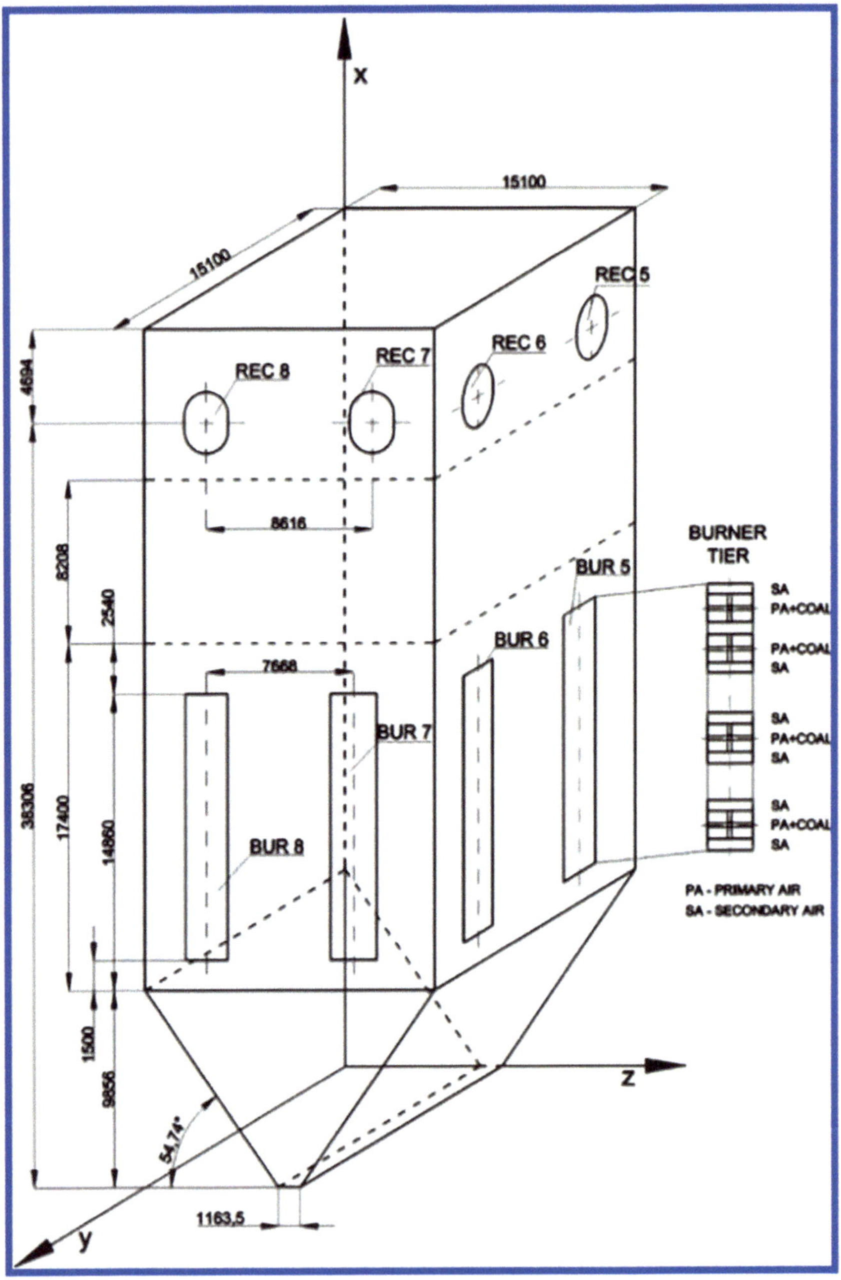

Fig. 3.104 Kostolac Power Plant B-1 and B-2 steam boilers furnace [60]

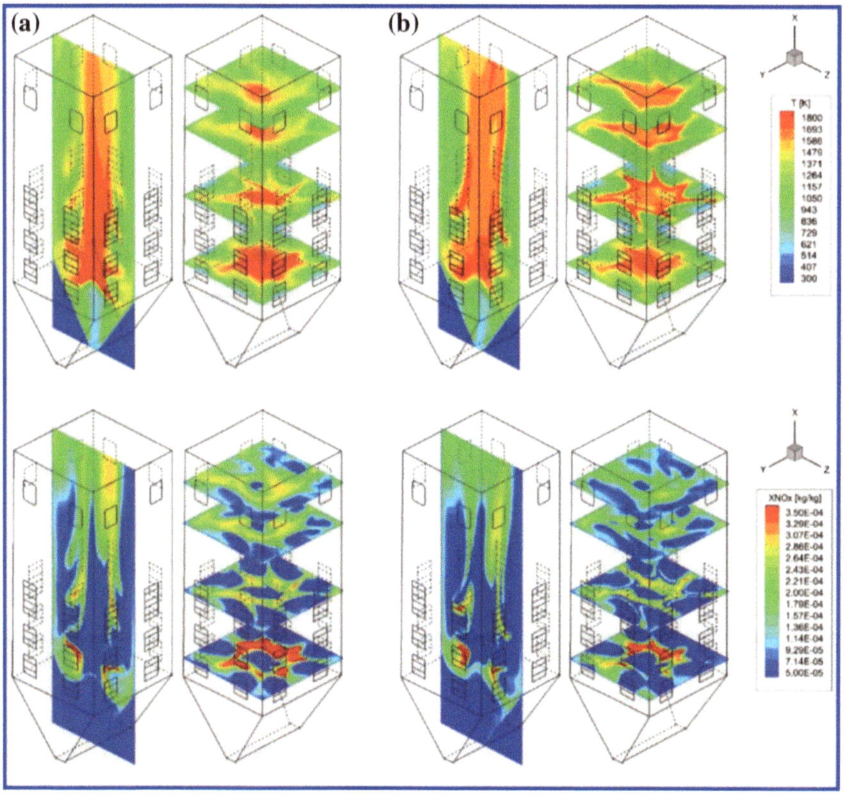

Fig. 3.105 Uneven distribution of fuel and air over the individual burners, temperature field, and the NO_x content in test cases 17 (**a**) and 28 (**b**): FEGT = 1015 and 993 °C, NO_x emission = 428.0 and 307.7 mg/Nm^3, respectively [60]

reduce fuel NO_x formation as well as thermal NO_x formation under the present simulated conditions. The decrease in the formation of fuel NO_x is due to the decreased contact of the nitrogen from the fuel with the oxygen within the combustion air, while the decrease in thermal NO_x formation is caused by a decrease in temperature (Figs. 3.108, 3.109, and 3.110).

Zhao F. Tian and Peter J. Witt from Commonwealth Scientific and Industrial Research Organization in Australia numerically studied the performance of a 375 MW brown coal-fired furnace in the Latrobe Valley, Australia [62]. Turbulence was modeled by the standard $k - \varepsilon$ model. The Lagrangian stochastic tracking model was applied to analyze the gas/particles flow field. Radiation was described using the discrete transfer model. Devolatilization was modeled with the single kinetic rate model. Gas-phase combustion was modeled using the species transport model with a one-step reaction mechanism, the turbulence-chemistry interaction using the eddy-dissipation model, and char combustion using the global reaction model. A level of confidence in the current CFD model has been established by

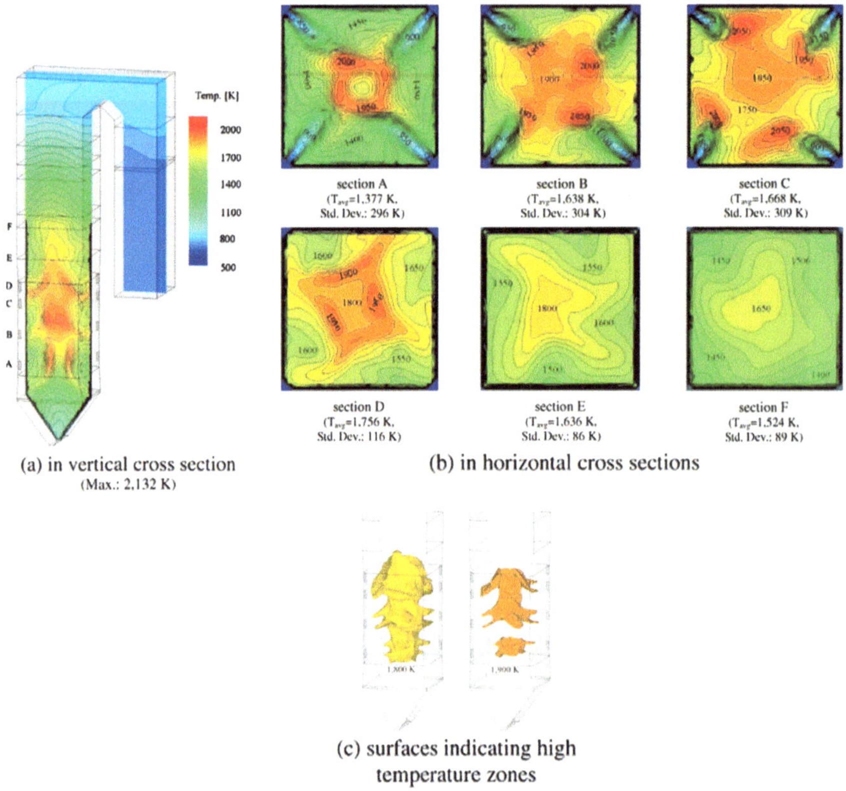

Fig. 3.108 Temperature distribution [61]

and mass change in Lagrangian field, with diffusion mode of particle dispersion by turbulence, while calculations of gas/particle two-phase coupling employed the particle-source-in-cell (PSIC) method. The six-flux method was used to model thermal radiation. Char combustion was modeled by employing a diffusion/kinetics model. The numerical predictions of the flame characteristics enable a specific tool for optimization of the boiler unit with respect to efficiency and ecology (Figs. 3.114, 3.115, 3.116, and 3.117).

Linbo Yan and Boshu He from Beijing Jiaotong University from China used CFD method to simulate the performances of different tangential arrangements of burners, including CTFB (corner tangentially fired boiler), WCTFB (wall-center tangentially fired boiler), and WOTFB (wall-off-center tangentially fired boiler), in the flow and combustion characteristics of pulverized coal in a 600 MW utility boiler [64]. A non-premixed combustion model is adopted by comparison, with the realizable $k - \varepsilon$ turbulence model to simulate the turbulent gas transport and discrete ordinates (DO) radiation model to calculate radiation heat transfer. The chemical reaction rate is determined by a simplified probability density function

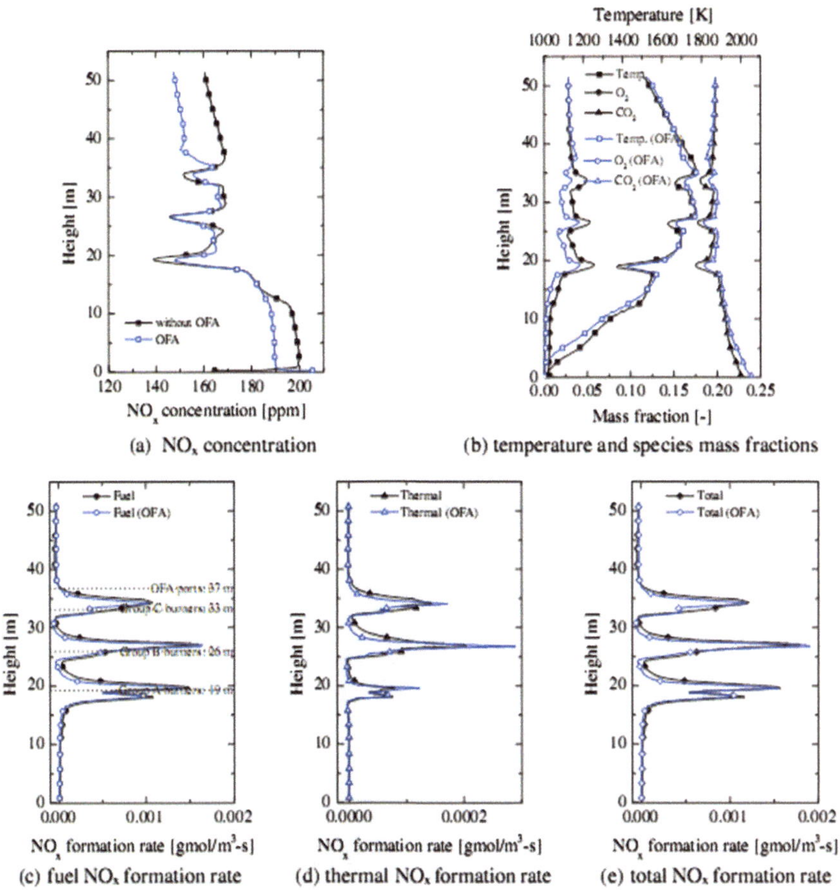

Fig. 3.109 Comparisons of results with and without OFA operation (average in each horizontal cross section along the furnace height) [61]

(PDF). Because of the variation in coal and inherent characteristics of the corner tangentially fired furnace, slagging and high-temperature corrosion on the surfaces of water walls and large temperature deviation at the furnace outlet are often detected during operation. The results show that, in contrast to the CTFB, the high-speed air flow erosion does not happen to the wall tangential burners. In the WCTFB, the high-temperature flame center is lowered, benefitting the burnout of coal particles, the decrease of the carbon content in fly ash, and the reduction of energy loss. In addition, combustion of coal in a local reducing atmosphere is avoided, which reduces the possibility of slagging and high-temperature corrosion. Because of the entrainment effects, uneven wall heating is eased. The flue gas temperature deviation and overheating at the furnace outlet are restrained, benefitted from the symmetric distributions of velocity and temperature (Figs. 3.118, 3.119, 3.120, 3.121, 3.122, and 3.123).

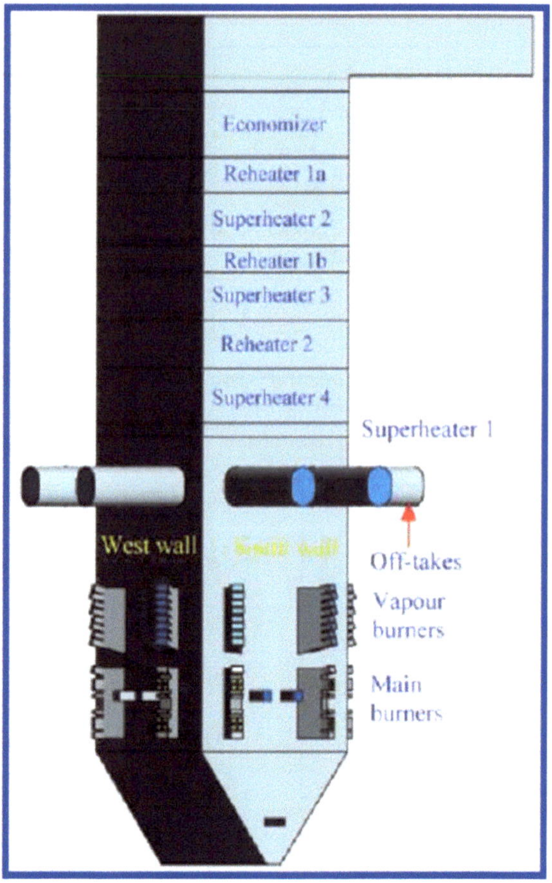

Fig. 3.110 Geometry of the CFD model for TRU energy Yallourn unit number 3 [62]

Hao Zhou and Guiyuan Mo from Zhejiang University in China carried out CFD investigations to understand the combustion and NO_x emission characteristics in a 1000 MW pulverized coal boiler, which is equipped with a dual-circular tangential-firing system in a single furnace [65]. In this work, a multigroup of SOFA nozzle arrangement was investigated for the NO_x control and reduction. Turbulence was modeled by the realizable $k - \varepsilon$ model. The coal particle motion was calculated by Newton's second law in the scheme of the Lagrangian method. The discrete ordinates radiation model was used to calculate radiation heat transfer. Devolatilization was modeled with the two-competing-rate model. The combustion of volatiles was modeled by employing the non-premixed combustion model with a PDF theory and char combustion by employing the diffusion/kinetics model. The formation of NO_x only includes thermal NO and fuel NO. The flow field, temperature distribution, species concentration, and char burnout are discussed. The relationship between the NO_x formation and the SOFA arrangement is analyzed. The numerical results show that the multigroup of SOFA nozzle arrangement is an

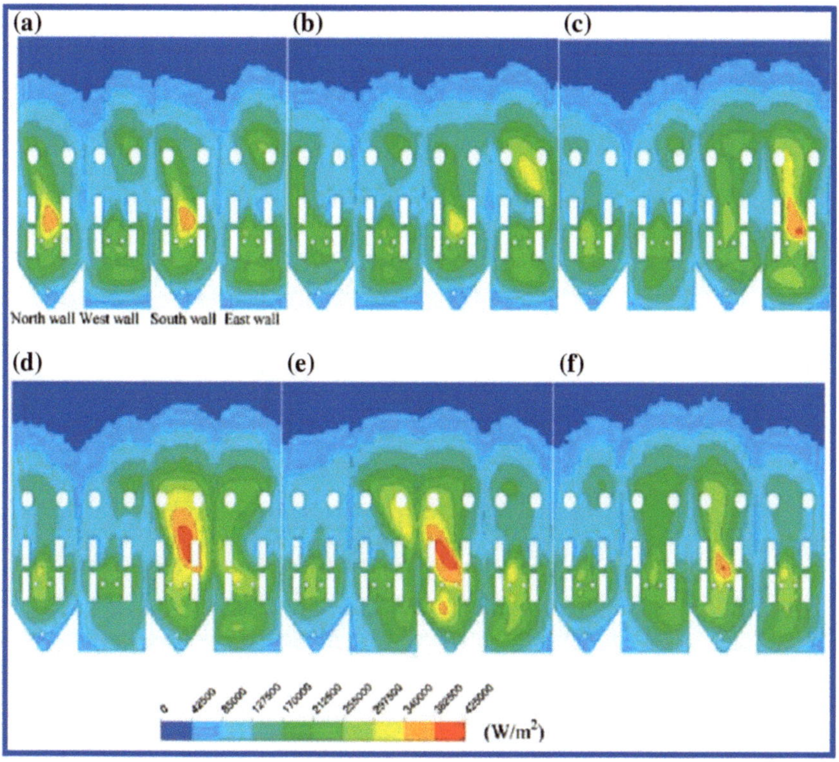

Fig. 3.111 Predicted wall incident heat flux: **a** case 1, **b** case 2, **c** case 3, **d** case 4, **e** case 5, and **f** case 6 [62]

efficient method to control the NO_x emissions in the pulverized coal boiler, with few unfavorable effects on the coal burnout. The arrangement of the two-group of SOFA nozzles being in service presents better ability than the one-group arrangement in reducing the NO_x emissions. The distance between the two groups of the SOFA nozzles should be longer for a better result of NO_x reduction (Figs. 3.124, 3.125, 3.126, and 3.127).

Feng Ren and Zhengqi Li from Harbin Institute of Technology in China developed a CFD model of a 300 MWe down-fired furnace equipped with overfire air (OFA) to investigate the effects of several operating conditions at full load with different OFA nozzle angles [66]. Turbulence was modeled by the standard $k-\varepsilon$ model. The Lagrangian stochastic tracking model was applied to analyze the gas/particles flow field, while calculations of gas/particle two-phase coupling employed the particle-source-in-cell (PSIC) method. Radiation was modeled using the P-1 model. Devolatilization was modeled with the two-competing-rate model. The combustion of volatiles was modeled by employing a probability density function theory and char combustion by employing the diffusion/kinetics model. The formation of NO_x only includes thermal NO and fuel NO. In the down-fired furnace,

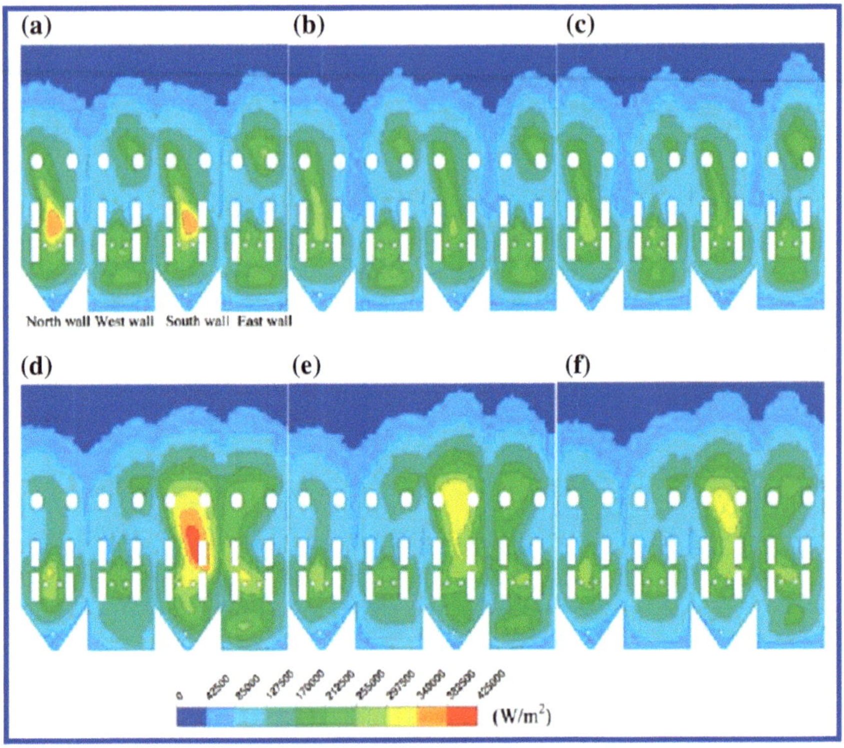

Fig. 3.112 Predicted wall incident heat flux: **a** case 1, **b** case 7, **c** case 9, **d** case 4, **e** case 8, and **f** case 10 [62]

the primary air/fuel was found unable to penetrate the horizontal secondary air flow zone and reach the furnace hopper; thus, no combustion occurs in the furnace hopper zone. The peak-temperature zone appears in the upper furnace, contrary to the original design concept. Simulation results also indicate that an OFA nozzle angle set below 30° moves the mixing point of the OFA flow and the up-flowing gas downward, further aiding combustion in the upper furnace. However, when the angle increases above 30°, OFA is unable to reach the furnace center, weakening combustion in the furnace, because no reaction with coal can take place (Figs. 3.128, 3.129, and 3.130).

Qingyan Fang and Huajian Wang from Huazhong University of Science and Technology in China systematically studied improving the performance of a 300 MW down-fired pulverized coal utility boiler by inclining downward the F-lay secondary air (SA) [67]. A numerical method was adopted to evaluate the effects of inclined angles on the characteristics of flow, combustion, and nitrogen oxide (NO_x) emissions in the furnace. Turbulence was modeled by the standard $k - \varepsilon$ model. The Lagrangian stochastic tracking model was applied to analyze the gas/particles flow field. Radiation was modeled using the P-1 radiation model. Devolatilization

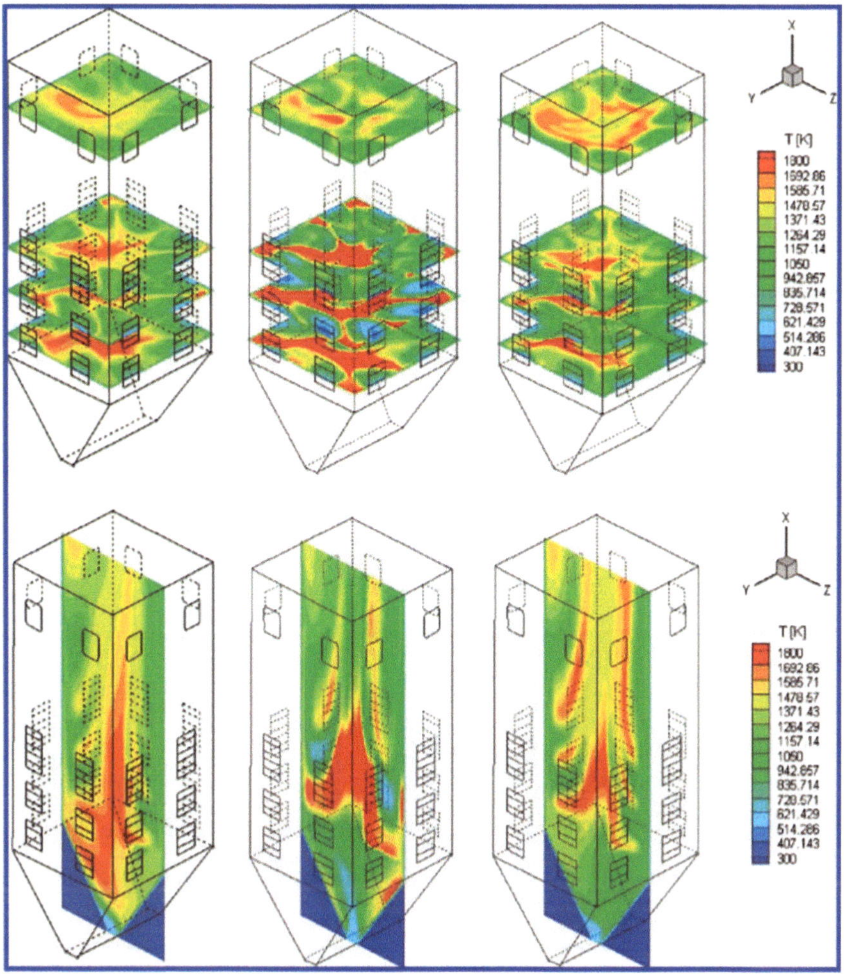

Fig. 3.113 Influence of the fuel and air distribution over the burner tiers on the flame geometry and position in test cases 1–3 [63]

was modeled with the two-competing-rate model. The combustion of volatiles was modeled by employing the non-premixed combustion model with a PDF theory and char combustion by employing the diffusion/kinetics model. The formation of NO_x only includes thermal NO and fuel NO. Retrofitting was conducted to incline the F-layer SA downward with an optimal inclined angle of 25°. The full-scale experimental measurements were carried out before and after retrofitting. The results indicate that inclining downward the F-layer SA can increase the flame penetration depth and lower downward the flame center. The residence time of pulverized coal particles increases, and the SA staging level enhances in the furnace. The boiler performance is improved with absolute increases of 3.55, 3.31, and

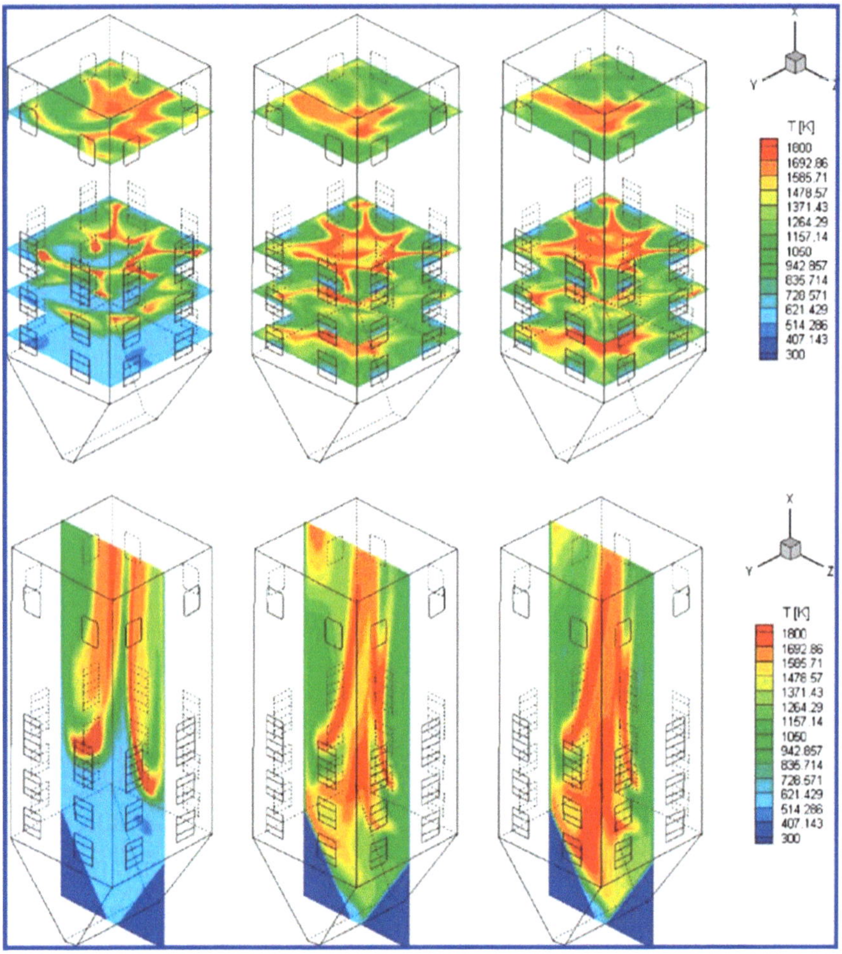

Fig. 3.114 Influence of the fuel and air distribution over the burner tiers on the flame geometry and position in test cases 4–6 [63]

2.20% in boiler efficiencies and relative reductions of 28.65, 19.07, and 12.53% in NO_x emissions under 300, 240, and 190 MW loads, respectively (Figs. 3.131, 3.132, 3.133, 3.134, and 3.135).

Javier Pallarés and Antonia Gil from University of Zaragoza in Spain numerically studied the cofiring coal and Cynara in a 350 MWe utility boiler. The most influential operational factors related to the biomass feeding conditions such as biomass mean particle size, level of substitution of coal by biomass, and feeding location in the furnace are analyzed, determining their influence in the combustion process. Turbulence was modeled by the standard $k-\varepsilon$ model. The Lagrangian stochastic tracking model was applied to analyze the gas/particles flow field, while calculations of gas/particle two-phase coupling employed the particle-source-in-cell

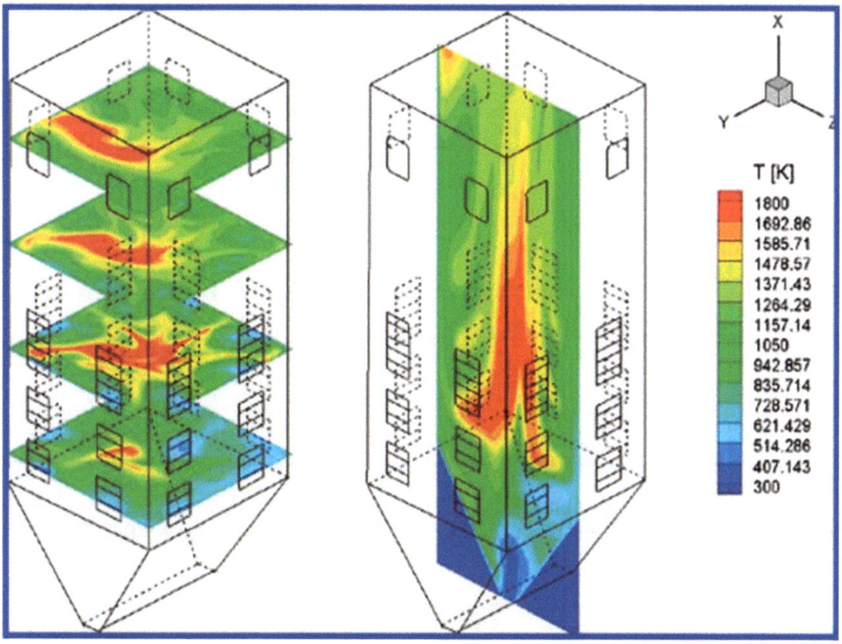

Fig. 3.115 Influence of the cold air ingress in test case 7 on the flame geometry and position [63]

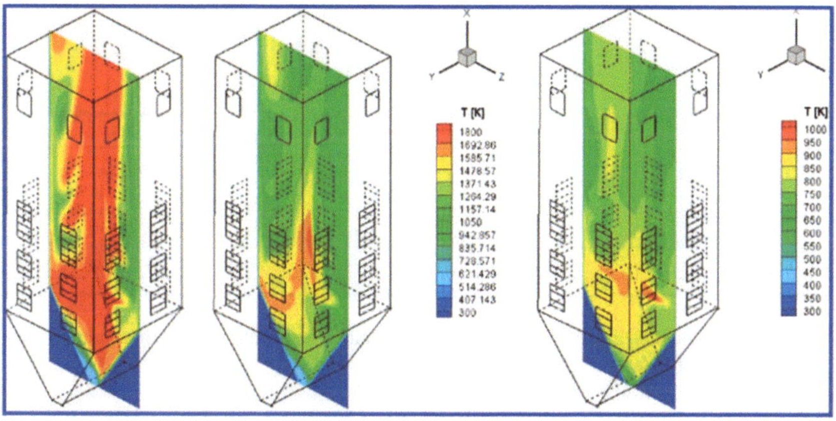

Fig. 3.116 Influence of the ash content in the pulverized coal on the flame geometry and position in test cases 8–10 [63]

(PSIC) method. Radiation was modeled using the P-1 model. Devolatilization was modeled with the single kinetic rate model. The gas-phase homogeneous combustion was modeled by employing the non-premixed combustion model with a PDF theory, and char heterogeneous combustion was modeled by employing the

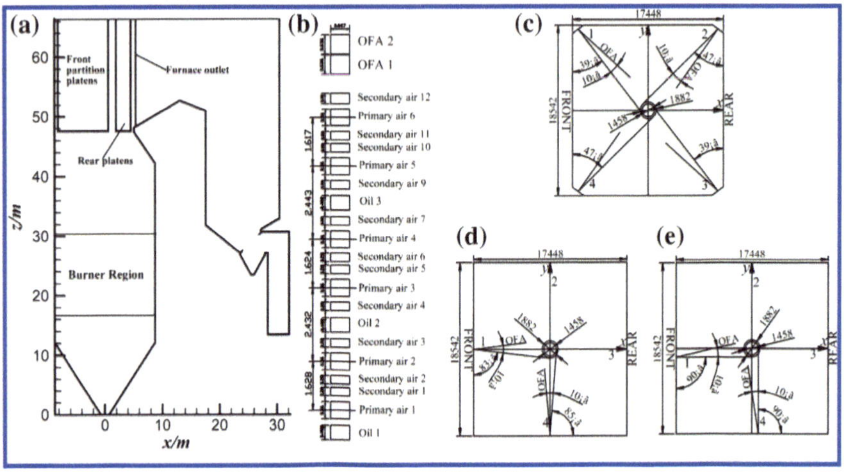

Fig. 3.117 Schematic of the furnace and the arrangements of the burners: **a** right view of the furnace, **b** nozzle arrangement for a burner set, **c** CTFB, **d** WCTFB, and **e** WOTFB [64]

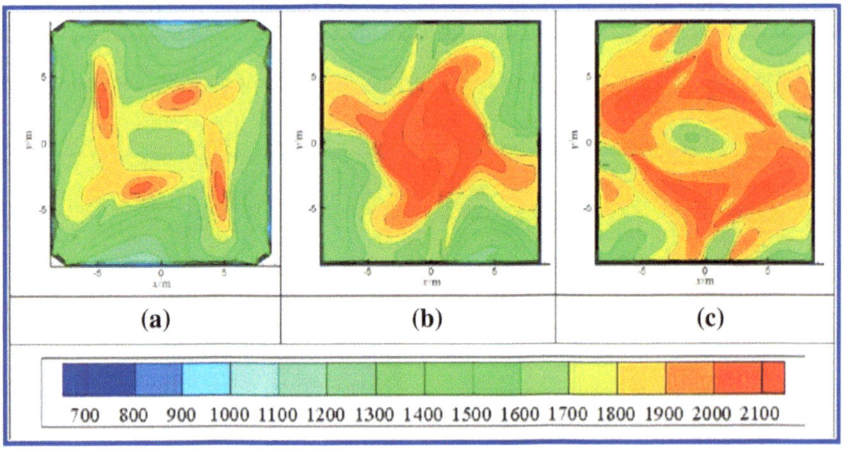

Fig. 3.118 Contours of the temperature on the horizontal cross section of the first primary air of the burners: **a** CTFB, **b** WCTFB, and **c** WOTFB [64]

single-film model. Particle size has turned to be the most influential factor in biomass combustion efficiency. Typical values in the range of 0.5–1 mm should be achieved during the milling process in order to obtain an adequate burnout while maintaining the mill consumption into tolerable values. A reduction in the combustion efficiency was obtained as long as the biomass portion was increased. However, this efficiency reduction maybe cut down if specific biomass burners were installed, instead of keeping the original coal ones. Residence times for the biomass particles are higher when biomass is fed at a lower intermediate row.

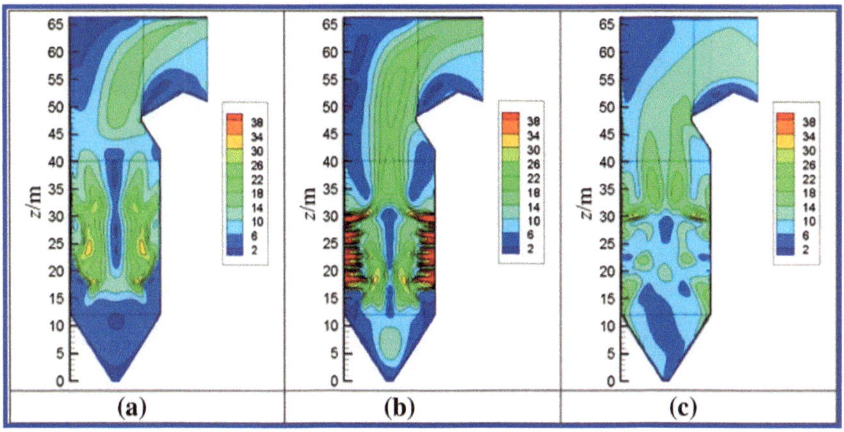

Fig. 3.119 Contours of the velocity on the central vertical cross sections for the furnaces at $y = 0$ m: **a** CTFB, **b** WCTFB, and **c** WOTFB [64]

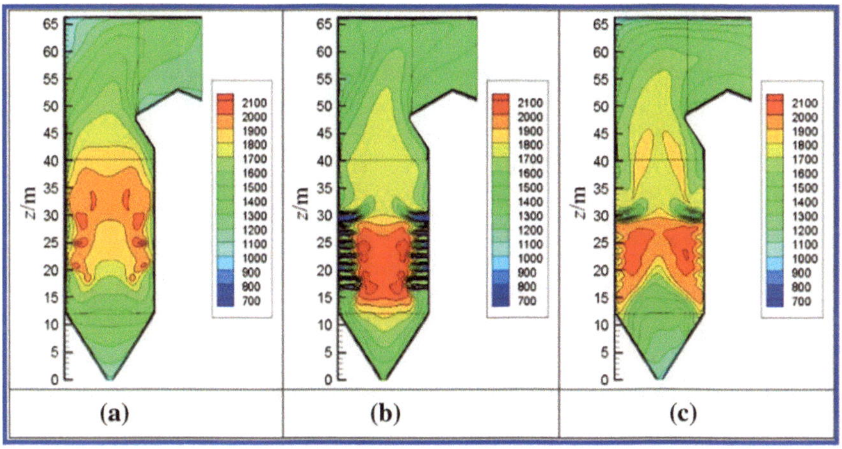

Fig. 3.120 Contours of the temperature on the central vertical cross sections for the furnaces at $y = 0$ m: **a** CTFB, **b** WCTFB, and **c** WOTFB [64]

However, other aspects, related to the specific nature of Cynara and the use of a nonspecific biomass burner, modify the oxygen competence between coal and biomass, the fluid-dynamic behavior, and the temperature–oxygen patterns in the boiler, thus affecting the combustion behavior as well. Results from the simulations have shown that the upper row location provides better combustion efficiency in the boiler (Figs. 3.136 and 3.137).

Michalis Agraniotis and Dimitris Stamatis from National Technical University of Athens in Greece investigated the changes in the combustion behavior caused by firing predried lignite in a 75 MWth lignite-fired Greek boiler through experimental

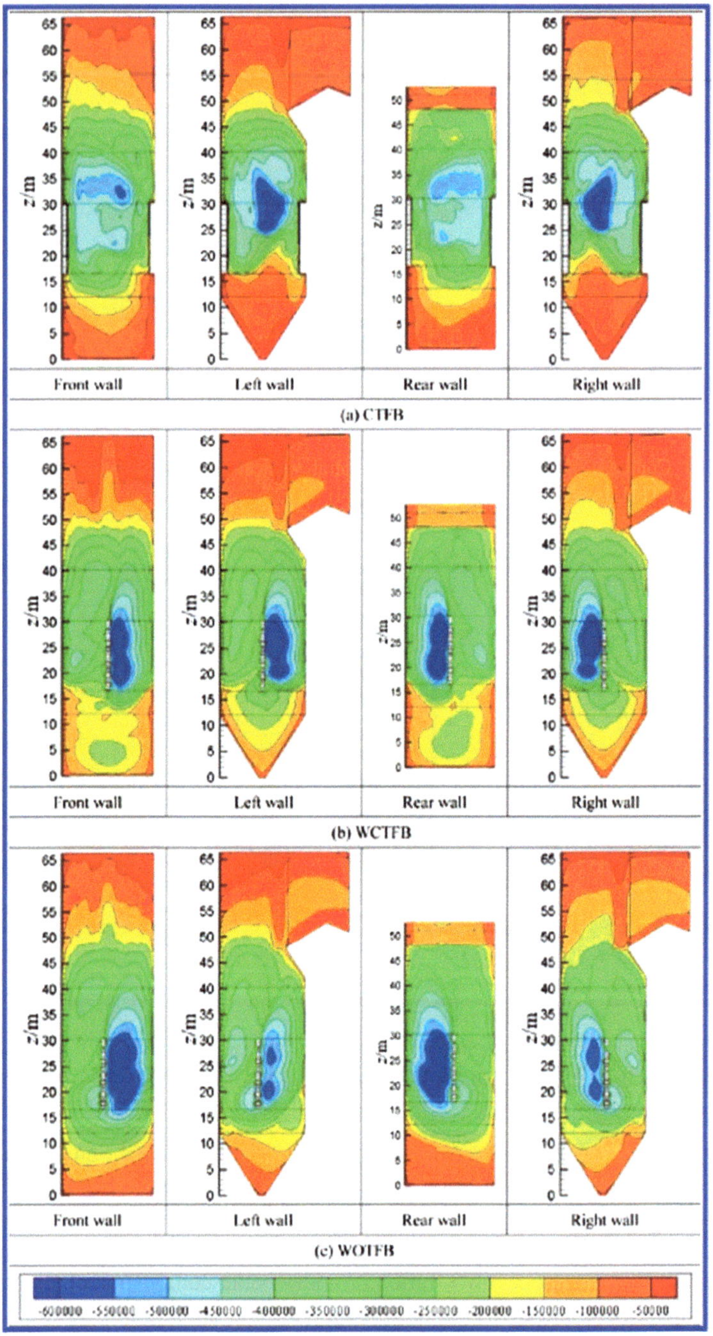

Fig. 3.121 Contours of heat flux (W/m^2) on the walls: **a** CTFB, **b** WCTFB, and **c** WOTFB [64]

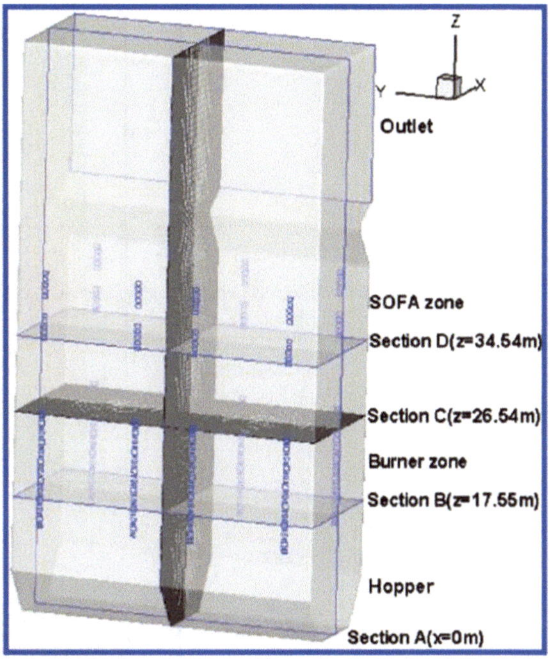

Fig. 3.122 Schematic configuration of the 1000 MW pulverized coal boiler

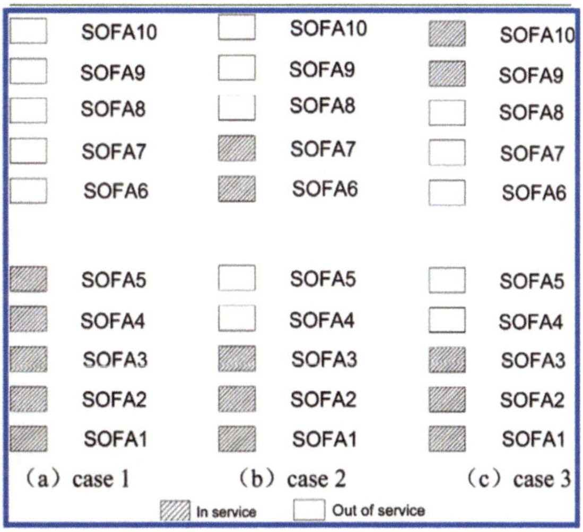

Fig. 3.123 SOFA nozzle arrangement for three cases

activities and numerical simulations [69]. The specific unit is equipped with dedicated dry lignite burners. Turbulence was modeled by the realizable $k-\varepsilon$ model. Discrete phase trajectories were calculated using Lagrangian formulation, and the coupling between the phases was introduced through particle sources of Eulerian gas-phase equations. The P-1 radiation model was used for the calculation of the

Fig. 3.124 Temperature distributions [65]

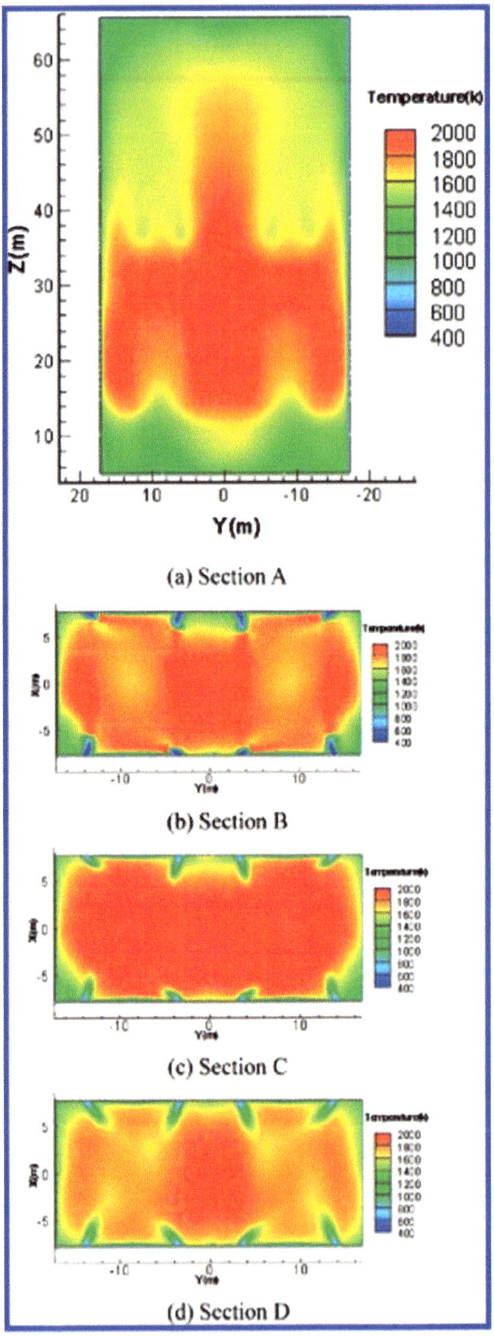

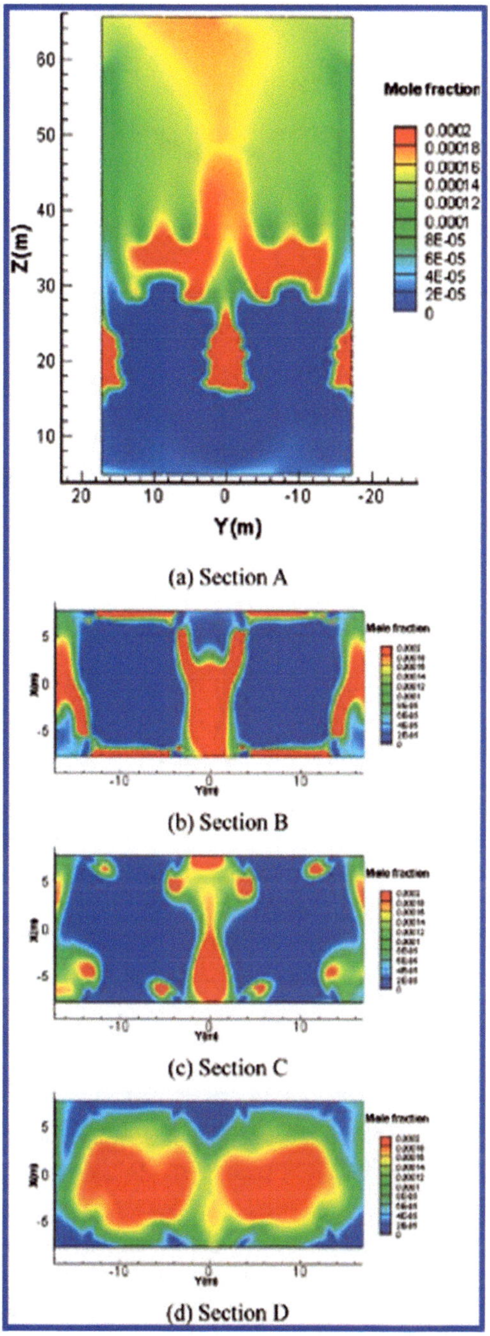

Fig. 3.125 Mole fraction distributions of NO_x [65]

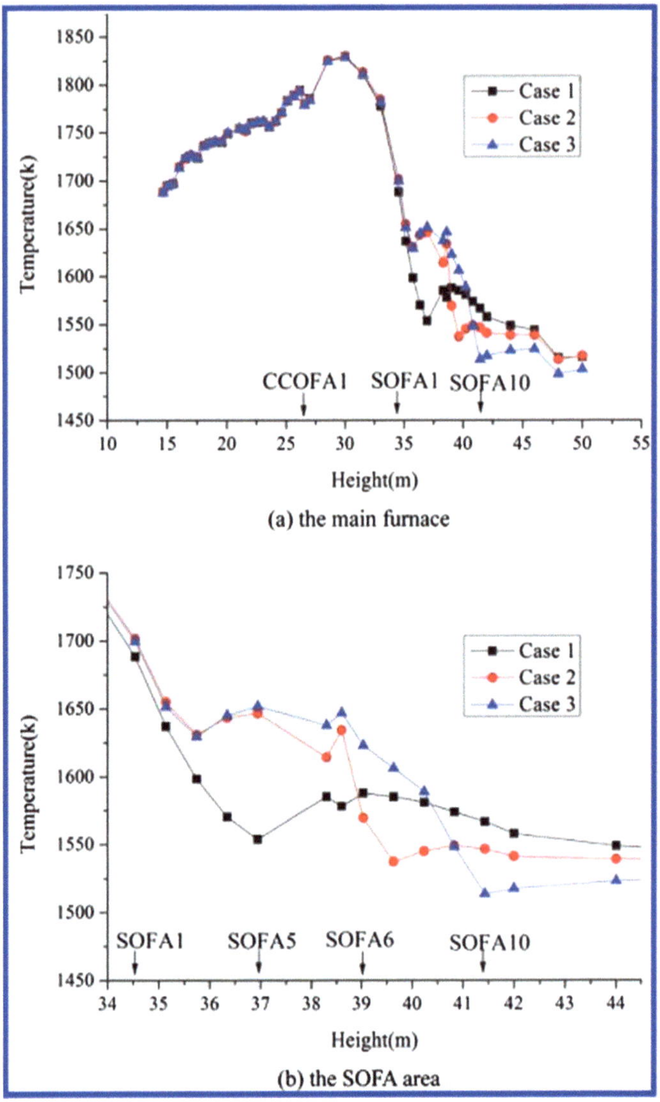

Fig. 3.126 Average temperature in each horizontal cross section along the furnace height for three cases [65]

radiative heat flux. Devolatilization was modeled with the single kinetic rate model. Gas-phase combustion was modeled using the species transport model with a two-step reaction mechanism, the turbulence-chemistry interaction using the eddy-breakup (EBU), and char combustion using the diffusion/kinetics model. The formation of NO_x only includes thermal NO and fuel NO. The measurements are performed with a dry lignite thermal share of 6%. Higher cofiring thermal shares of

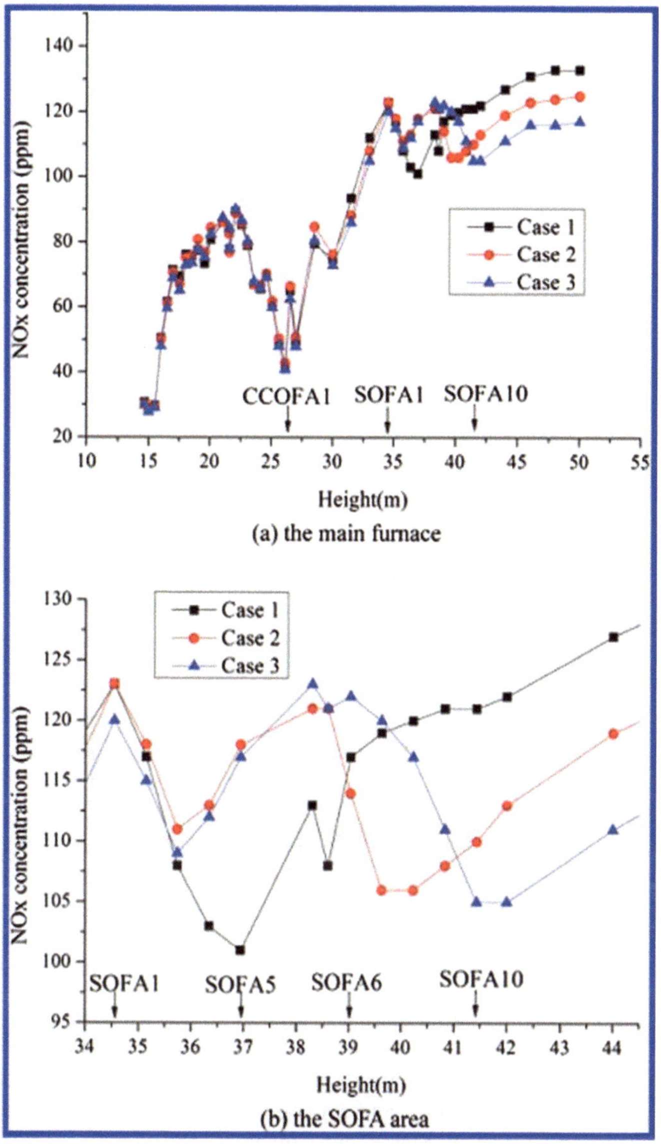

Fig. 3.127 Average NO_x concentration in each horizontal cross section along the furnace height for three cases [65]

up to 20% are further simulated and the effect of cofiring on the combustion behavior is evaluated by specific parameters including temperature fields, wall heat flux, fuel's burnout, and NO_x emissions. Some trends are observed in the simulations of the dry coal cofiring cases. Increased temperature peaks in the near burner region, and higher furnace outlet temperatures are predicted. A clear increase of the

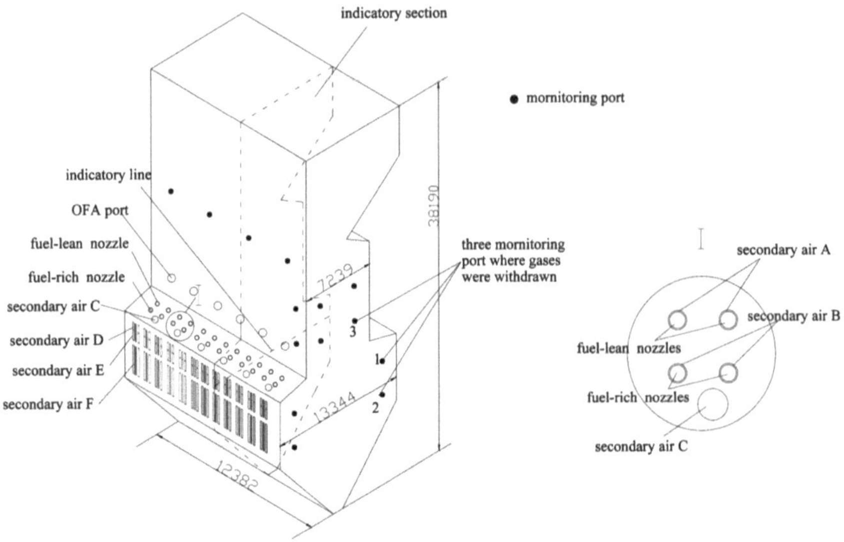

Fig. 3.128 Geometry of the CFD model for the FW-type 300 MWe down-fired boiler [66]

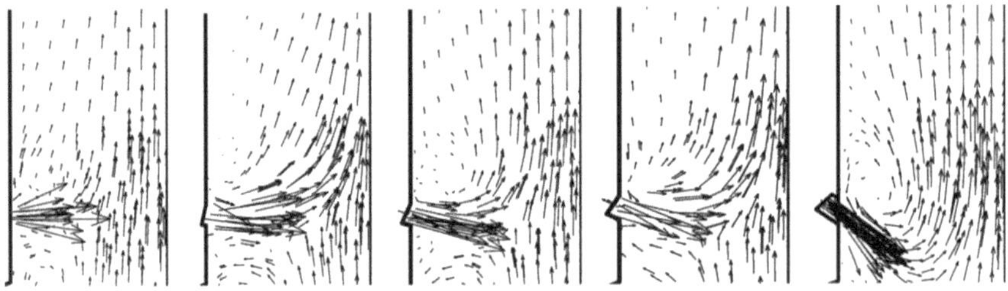

Fig. 3.129 Flow fields at the outlet of an OFA nozzle with different nozzle angles [66]

total fuel burnout when firing dry coal is also foreseen by the simulations, which is a strong argument for the coutilization of dry lignite as a supportive fuel (Figs. 3.138, 3.139, 3.140, 3.141, and 3.142).

Taha J. Taha and Arthur F. Stam from University of Twente in the Netherlands developed a CFD model to investigate the slagging behavior of meat and bone meal (MBM) at higher cofiring rates in a tangentially fired boiler [70]. On the basis of the thermal heat input, solid fuel combustion is modeled and evaluated for various cofiring rates which consist of 0, 12.5, 25, and 40% of MBM. Turbulence was modeled by the realizable $k - \varepsilon$ model. Discrete phase trajectories were calculated using Lagrangian formulation, and the coupling between the phases was introduced through particle sources of Eulerian gas-phase equations. The discrete transfer radiation model theory was used to evaluate radiative heat transfer. Devolatilization

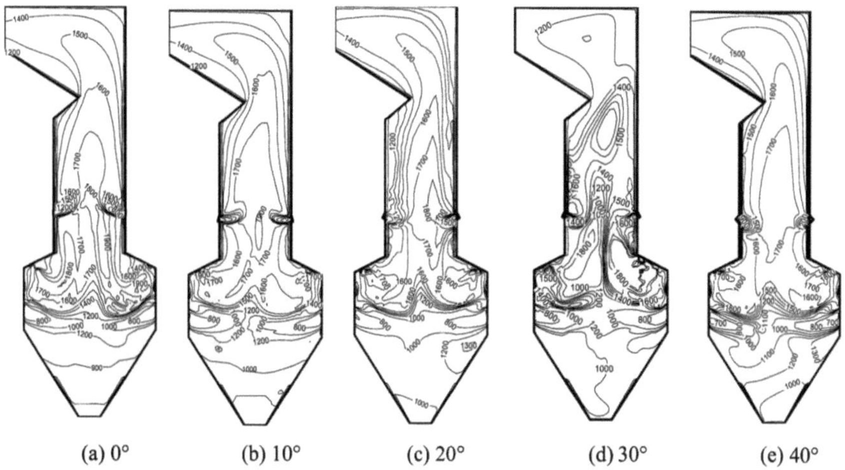

Fig. 3.130 Calculated gas temperature (K) over the indicatory section with different OFA nozzle angles [66]

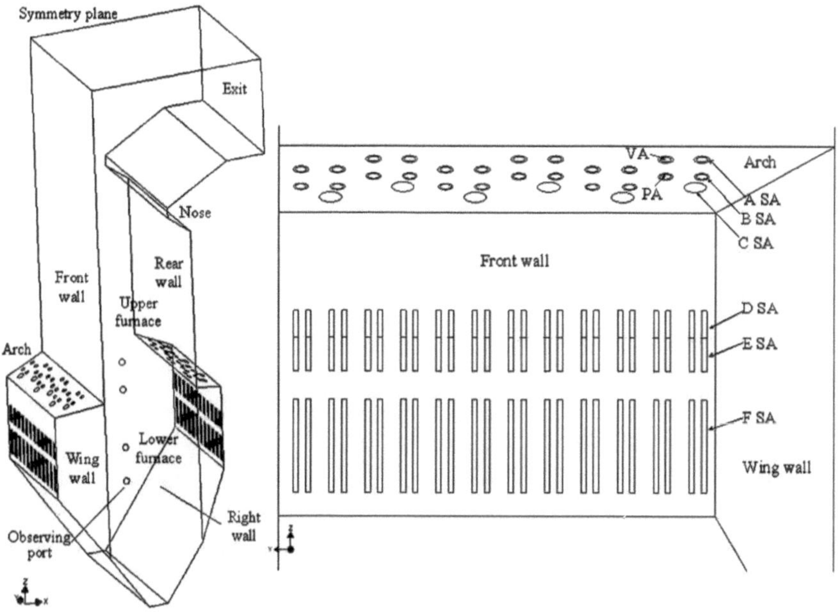

Fig. 3.131 Schematic diagram of the boiler (only half of the boiler is shown in the figure because the furnace structure is symmetrical) (PA, primary air; VA, vent air; and SA, secondary air) [67]

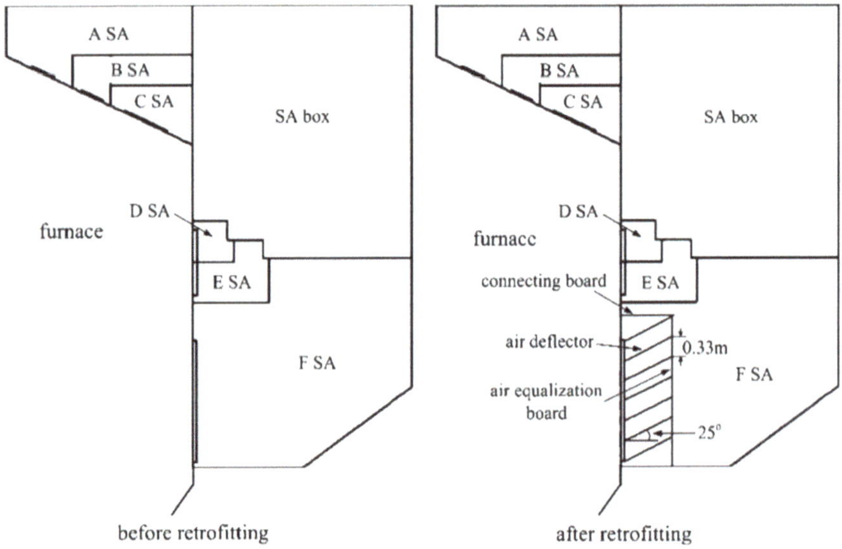

Fig. 3.132 Schematic diagram of the retrofit for F-layer SA [67]

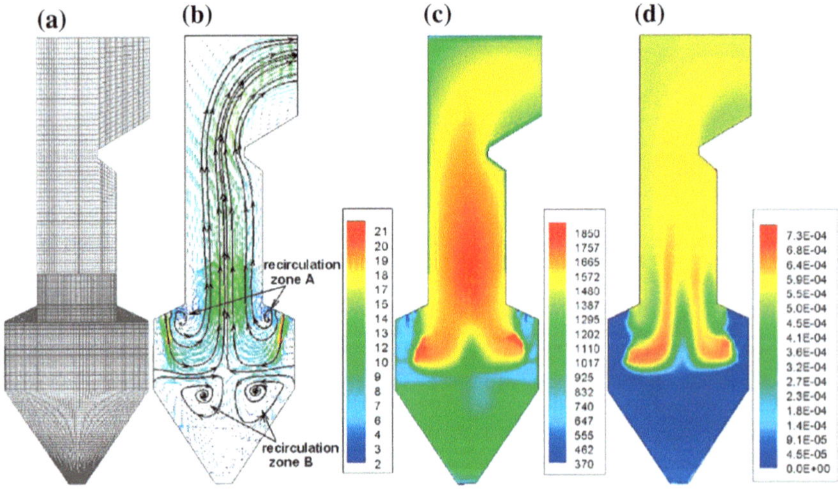

Fig. 3.133 Central cross section of the burner nozzle with horizontal F-layer SA: **a** grid used for computed results, **b** flow field (m/s), **c** temperature distribution (K), and **d** NO concentration distribution (106 ppm) [67]

was modeled with the single kinetic rate model. Gas-phase combustion was modeled using the species transport model with a two-step reaction mechanism, the turbulence-chemistry interaction using the eddy-dissipation model, and char combustion using the diffusion/kinetics model. An inertial impaction mechanism of fly

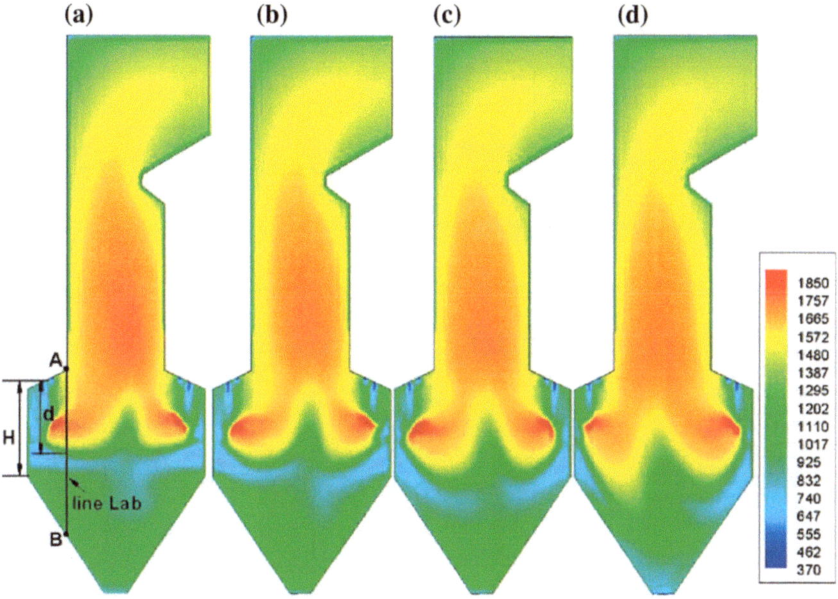

Fig. 3.134 Comparison of the temperature distributions (K) for the different inclined angles: **a** 0°, **b** 15°, **c** 25°, and **d** 35° [67]

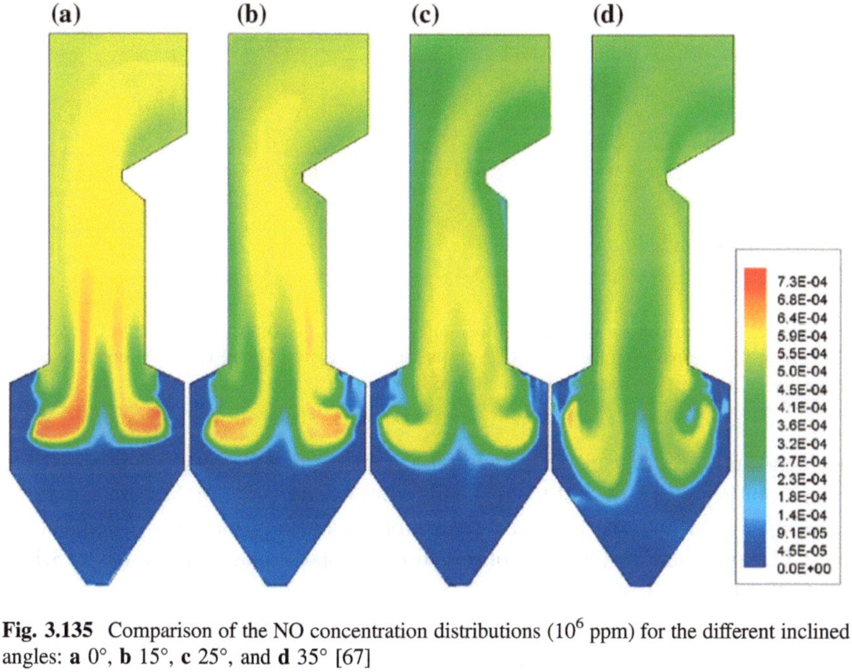

Fig. 3.135 Comparison of the NO concentration distributions (10^6 ppm) for the different inclined angles: **a** 0°, **b** 15°, **c** 25°, and **d** 35° [67]

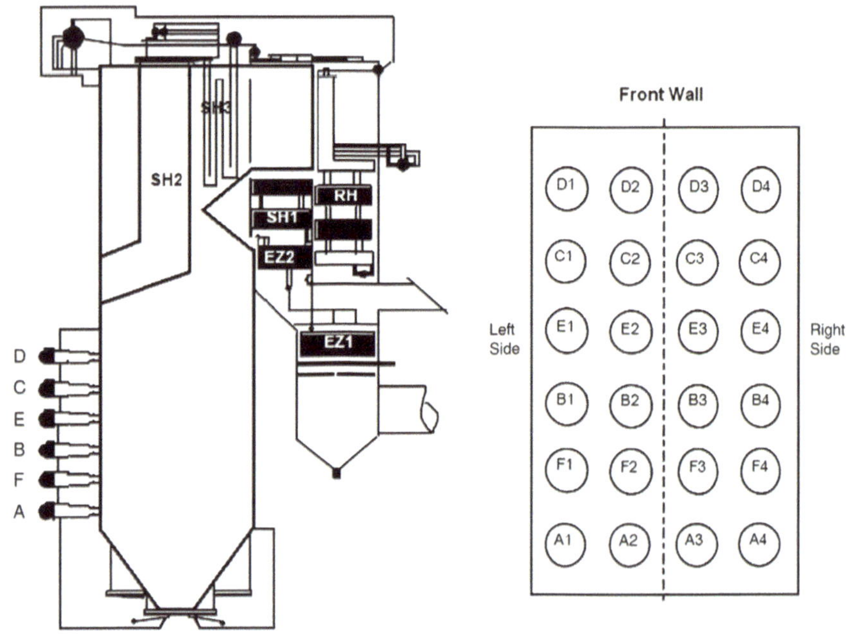

Fig. 3.136 Schematic layout of the Teruel power plant. Detail of burner's configuration [68]

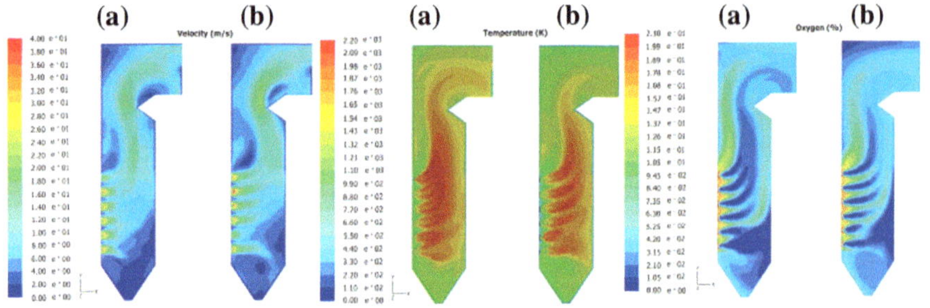

Fig. 3.137 Comparison of velocity, temperature, and oxygen contours plots at a cross section containing the biomass burner depending on its location: **a** row D and **b** row F; corresponding, respectively, to the simulation cases 6 and 7 [68]

ash deposition was considered only to predict the water walls slagging process. The calculation results show that the deposition propensity is the highest for a cofiring ratio of 25% MBM. The preferred deposition locations in the boiler calculated by the CFD model are in line with observations in operational practice (Figs. 3.143, 3.144, 3.145, and 3.146).

Audai Hussein AI-Abbas and Jamal Naser from Swinburne University of Technology in Australia developed a CFD modeling to investigate the brown coal combustion in a large-scale tangentially-fired furnace (550 MW) under different

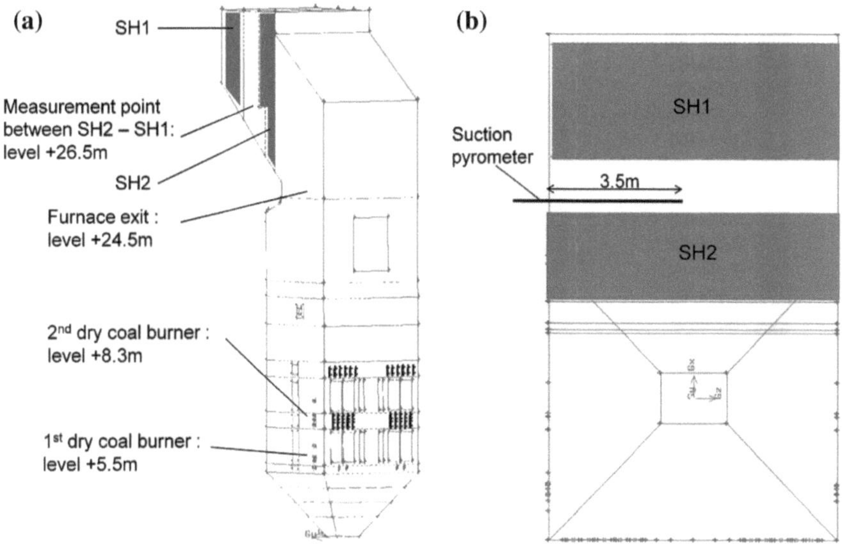

Fig. 3.138 **a** Drawing of the Liptol boiler and **b** boiler cross section at level 26.5 m

operating conditions [71]. The strategy of operation schemes for the first six combustion scenarios was based on the change of the out of service (turned off) burners under full-load operation, while the rest cases were carried out at 20% lower and 20% higher loads than the standard operating conditions. Turbulence was modeled by the standard $k - \varepsilon$ model. The discrete droplet method was used for the solid-phase flow simulation, and the discrete transfer radiation method for the radiative heat transfer calculations. Gas-phase combustion was modeled using the species transport model with a three-step global mechanism, the turbulence-chemistry interaction using the eddy-breakup (EBU) model, and char combustion using a model with three heterogeneous chemical reactions where the oxidation rate of the char is governed by the diffusion of bulk oxygen partial pressure to the particle's external surface and by the effective char reactivity. The validated model was used to perform the following investigation parameters: furnace gas temperatures, species concentrations (O_2, CO, and CO_2), velocity distributions, and char consumption. The predictions demonstrated that there are good temperature distributions in the furnace when the turned off burners are set in the opposite direction under full-load operation. For higher aerodynamic effect, the numerical results showed improvements on the combustion characteristics in terms of species concentrations and char burnout rates in comparison with the standard operating case (Figs. 3.147, 3.148, 3.149, and 3.150).

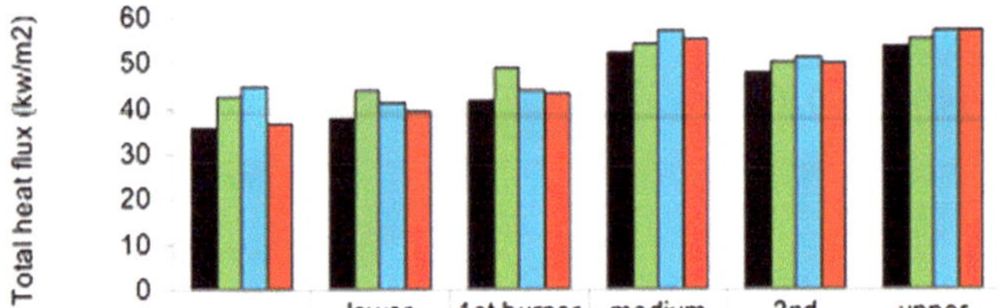

	hopper	lower zone	1st burner level	medium zone	2nd burner	upper zone
■ baseline	35,46	37,55	41,69	52,10	47,68	53,32
■ dry coal 5%	42,42	43,94	48,82	54,03	49,95	55,07
■ dry coal 10%	44,78	41,31	43,88	56,96	51,13	57,10
■ dry coal 20%	36,64	39,12	43,29	54,93	49,70	57,17

Fig. 3.142 Distribution of total heat flux (kilowatts/meters squared) on furnace zones [69]

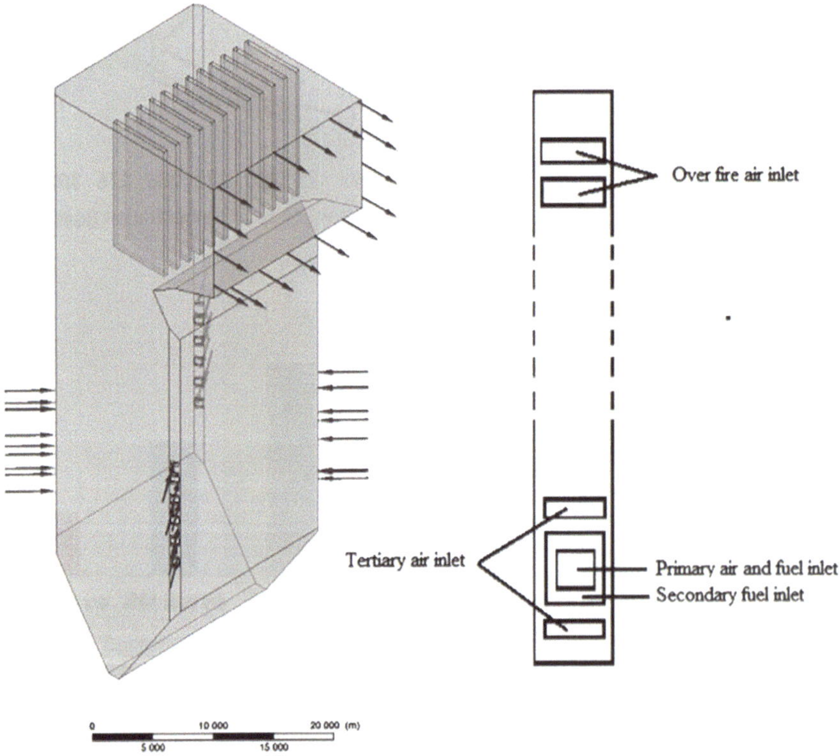

Fig. 3.143 Schematic representation of the boiler domain (left) and representative fuel and air inlet ports (right)

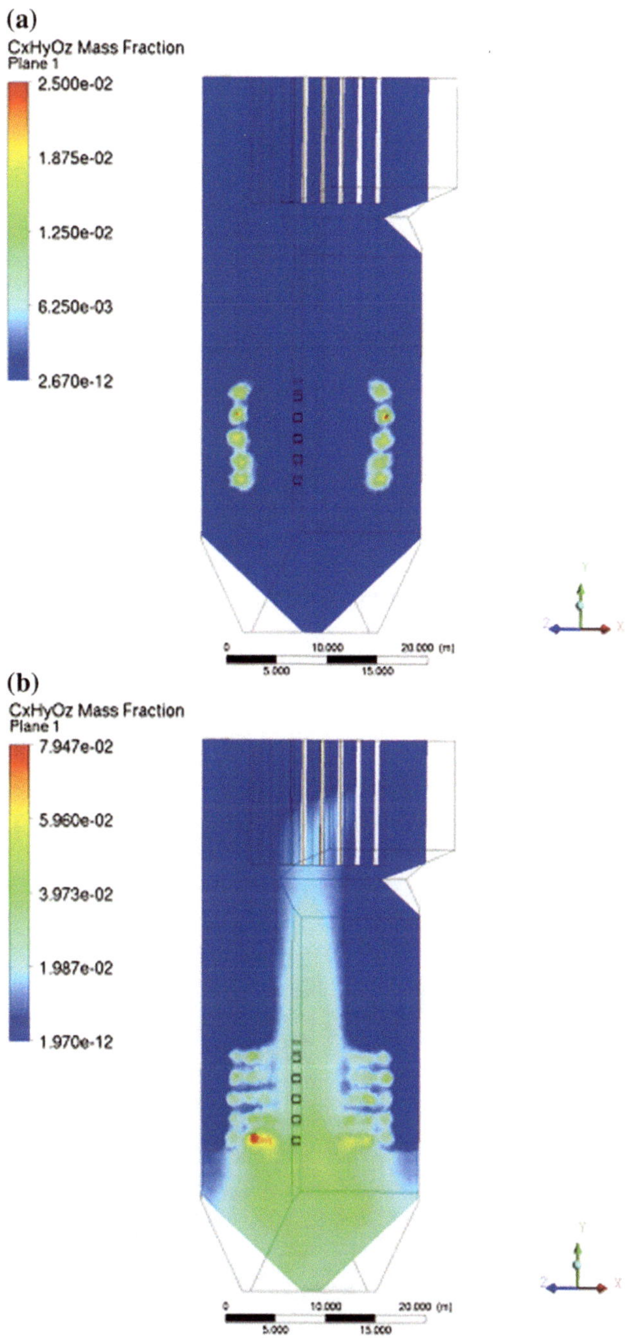

Fig. 3.144 Cumulative volatile mass fraction during **a** coal combustion and **b** cofiring 40% MBM with coal [70]

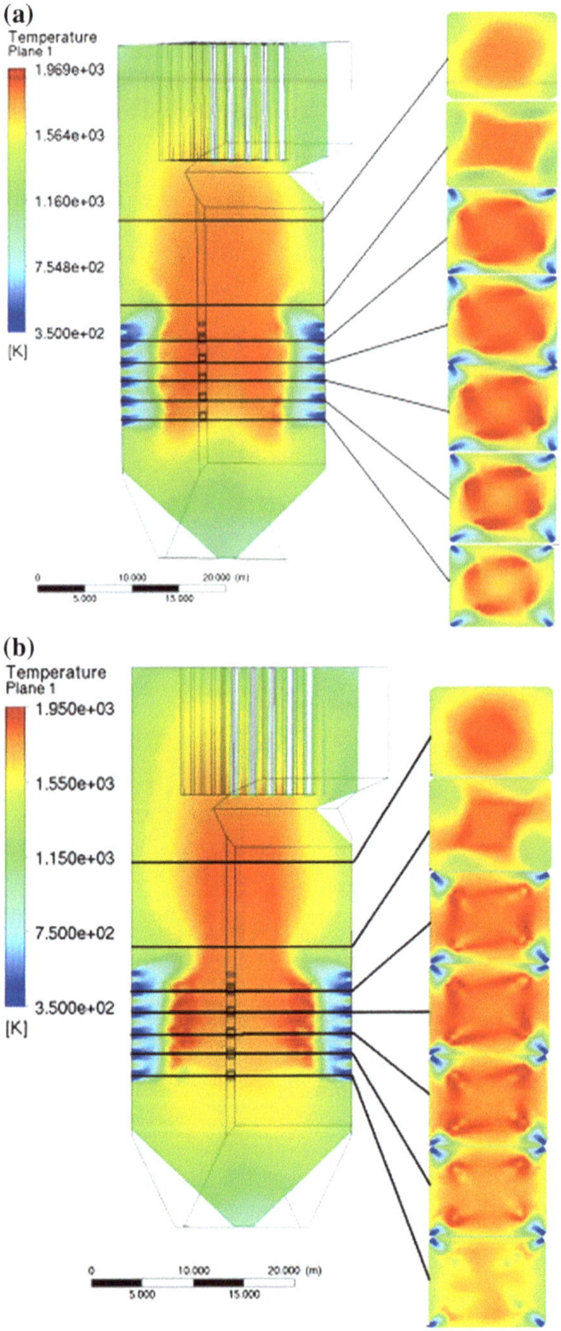

Fig. 3.145 Flue gas temperature profile along the boiler height (14, 16, 18, 20, 22, 26, and 35 m above the lower edge of the boiler) during **a** coal combustion and **b** cofiring of 40% MBM [70]

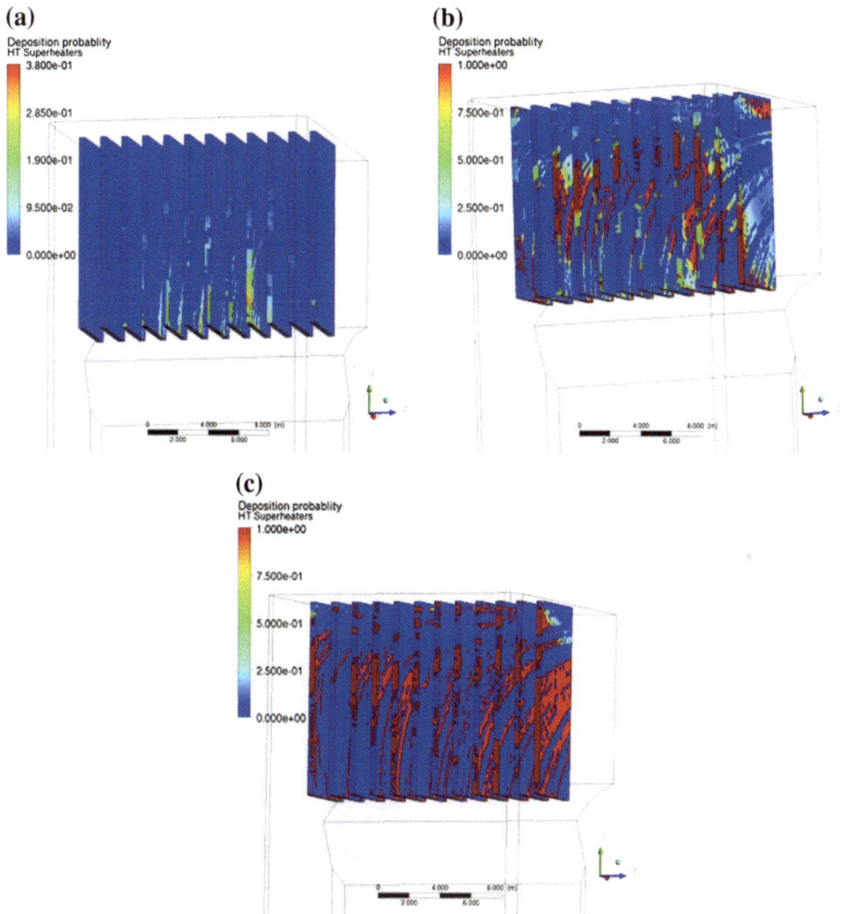

Fig. 3.146 Slagging propensity calculated using μ_{crit} = 8 Pa s for **a** 12.5%, **b** 25%, and **c** 40% MBM cofiring on the superheaters' surfaces [70]

multi-dimensionally and highly nonlinearly correlated to boiler operational parameters. It is difficult so far to establish and maintain a perfect model predicting NO_x emissions. Due to high nonlinearity, available approaches for NO_x emissions reduction are almost all empirical and require ample emission data. This implies a large number of experiments to be conducted during process operation. Fortunately, at present, the majority of industrial facilities are equipped with distributed control systems (DCS) and continuous emissions monitoring system (CEMS) which supply a data collection repository of operating parameters recorded in real time. Hence, empirical models such as black box models based on the daily operational records of power plants are feasible and very useful to modeling NO_x emissions. As indicated, optimizing boiler operation and emissions performance requires

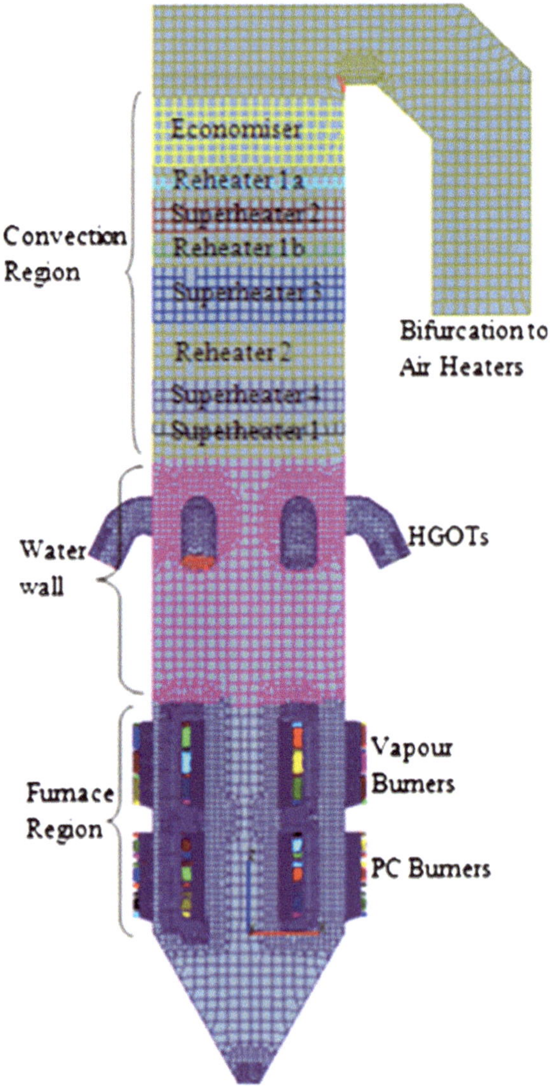

Fig. 3.147 CFD geometrical model of unit 1 at Loy Yang A power plant

embedding artificial intelligence or other advanced computer-based controls into a power plant's digital control system.

Artificial neural networks are well-known tools among artificial intelligence techniques, which are able to reproduce the relationships existing between input and output variables of highly nonlinear systems. Up to now, there are many research works that focused on applying neural networks to pollutants emissions modeling. GMDH-based abductive networks were employed to model mercury speciation in combustion flue gases. Fuzzy neural network was proposed to model NO_x emissions. Liu and Huang [72] proposed a fuzzy logic model to generate a

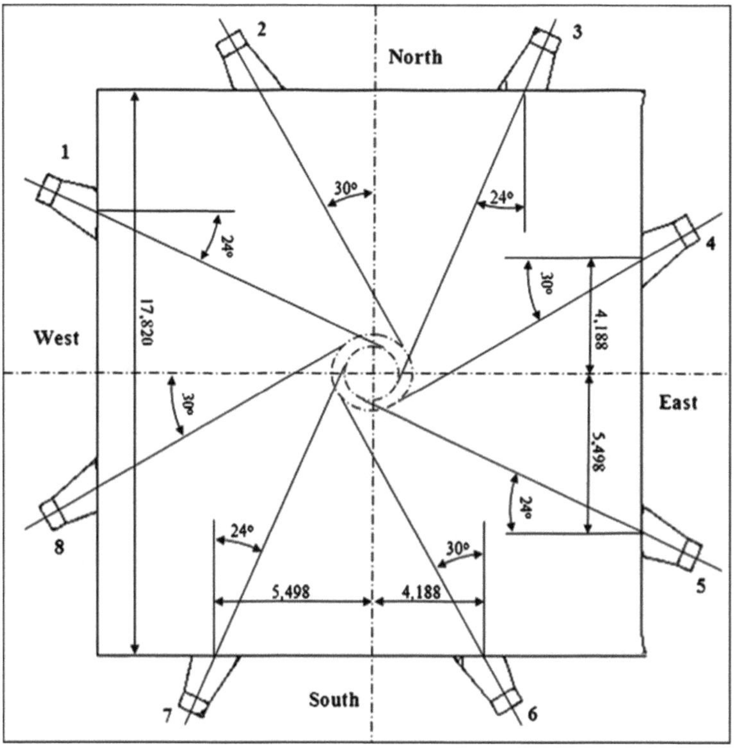

Fig. 3.148 The schematic representation of the burners' configurations

reliable emission model and deal with environmental and economic dispatching when only limited experimental data are available. An intelligent decision support system for air pollution control at coal-fired power plants was developed. Various neural networks, especially back propagation neural networks (BPNN), were developed for the modeling and control of the nitrogen oxide emissions. A time-delay ANN model was designed for the dynamic prediction of nitrogen oxides and carbon monoxide emissions from a fossil fuel power plant. Artificial neural networks (ANN) were employed to predict the nitrogen oxides (NO_x) emission characteristics of a large capacity pulverized coal-fired boiler. A genetic algorithm-based gray-box model was built to model and predict NO_x emissions in a 500 MWe coal-fired power plant. The pollution emissions monitoring approach using ANN-based software sensors had been developed. In summary, various variants of artificial neural networks for NO_x emissions modeling have attracted many researchers' attention in last ten years, as reviewed by Kalogirou [73]. However, the neural network suffers from a number of weaknesses, which include the needs for numerous controlling parameters, difficulty in obtaining a stable solution, and the danger of over-fitting.

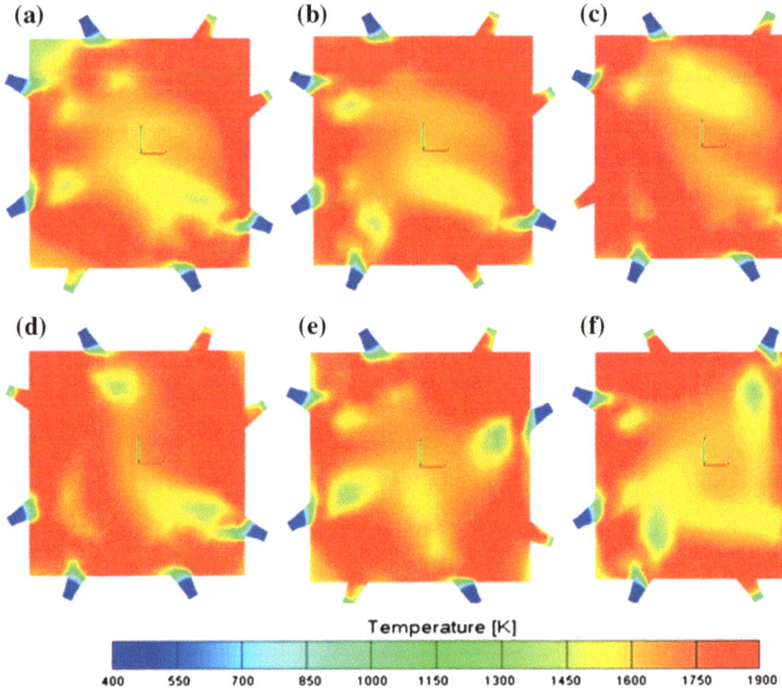

Fig. 3.149 Temperature distributions on the lower intermediate main burner at the cross section cuts (X–Y plane) for six combustion cases: **a** case 1, **b** case 2, **c** case 3, **d** case 4, **e** case 5, and **f** case 6

In recent years, SVR has been successful in mapping the complex and highly nonlinear relationship between system input and output. However, to the best of authors' knowledge, SVR has never been applied to model NO_x emissions of coal-fired utility boilers in the literature.

In the second stage of combustion modification, some optimization algorithms are used to manipulate the inputs of the model in order to minimize the emissions output. Obviously, the selection and the implementation of optimization algorithms are the main tasks of the second stage. To achieve an online and real-time application, optimization algorithms with quick convergence speed and good performance are highly desirable. Evolutional algorithms (EAs) are optimization techniques simulating biological systems which are classified as a category of the research of the so-called artificial life. Such algorithms have been developed to solve large-scale optimization problems, for which traditional mathematical techniques may fail. The typical EAs include genetic algorithm (GA), memetic algorithm, particle swarm optimization (PSO), ant colony systems (ACO), AbYSS, MOCell, OMOPSO, SPEA2, and shuffled frog leaping. Among these optimization

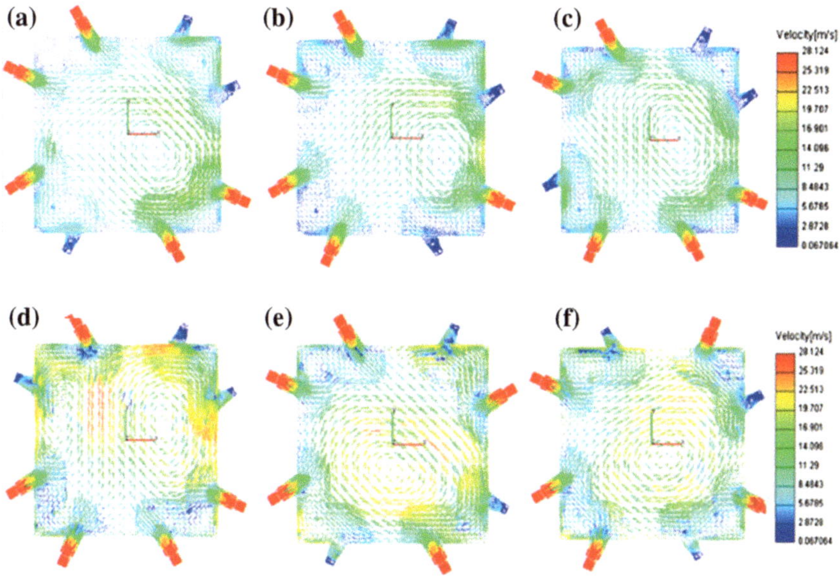

Fig. 3.150 Gas velocity vector on the upper intermediate inert of the secondary air duct for six combustion cases: **a** case 1, **b** case 2, **c** case 3, **d** case 4, **e** case 5, and **f** case 6

methods, only GA is used by few studies to optimize operational conditions of the boiler for NO_x emissions abatement. It is meaningful to introduce other optimization algorithms into the development of combustion optimization software package. However, there has hitherto not been any attempt at developing other optimization algorithms other than GA to reduce NO_x emissions. PSO and ACO have proved to be a promising optimization technology widely used in many science and engineering fields. In the present study, efforts are devoted to introduce these optimization techniques into combustion optimization for the utility boilers, which can provide an alternative approach to neural networks- and GA-based combustion optimization. AbYSS, MOCell, OMOPSO, and SPEA2 have believed to be the effective way to optimization multi-objects problems.

In next work, support vector regression and artificial neural network were proposed to model the relation between NO_x emissions and operational parameters of a utility boiler, as well as unburned carbon. Subsequently, GA, PSO, and two variants of ACO were employed to be combined with the SVR model of the NO_x emissions characteristics to find the optimal operating parameters which result in low NOx emissions and unburned carbon. The comparison on the availability and speed of the various optimization algorithms was conducted.

3.5 Summary

In this chapter, combustion characteristic modeling methods are introduced. Although combustion process is complex, certain character can be exposed by different studies. Experimental method gives detailed information about coal combustion, including ash, fusion temperature, flame temperature, and flue gas. Experimental method costs much but obtains accurate measurement. Different CFD methods including turbulence model, radiative heat transfer model, and so on and their application condition are introduced; CFD method is a convenient to get comprehensive particular information. Computational intelligence method is also discussed in this section, and it has been believed that computational intelligence is a suitable way to study combustion.

References

1. Lindsay J. What price coal quality in electric power generation. Aust Coal Rev. 1997;3.
2. Sun X. Coal-fired boiler combustion experimental techniques and methods. Beijing: China Power Press; 2001.
3. Xiong Y, Sun X. Analysis methods of coal combustion characteristic. Coal Qual Technol. 1998;5:27–31.
4. Congzhen F. Principles of boiler. Beijing: China Power Press; 1986.
5. Zhang K, Xia H, Deng X, et al. Test method for analysis of coal ash. Standards Press of China; 2008.
6. Umar DF, Usui H, Daulay B. Change of combustion characteristics of Indonesian low rank coal due to upgrade brown coal process. Fuel Process Technol. 2006;87(11):1007–11.
7. Cara J, Vargas M, Morán A, et al. Biodesulphurization of a coal by packed-column leaching. Simultaneous thermogravimetric and mass spectrometric analyses. Fuel. 2006;85(12–13): 1756–62.
8. Zhang C, Jiang X, Wei L. Research on pyrolysis characteristics and kinetics of super fine and conventional pulverized coal. Energy Convers Manag. 2007;48(3):797–802.
9. Sharypov VI, Beregovtsova NG, Kuznetsov BN, et al. Influence of reaction parameters on brown coal-polyolefinic plastic co-pyrolysis behavior. J Anal Appl Pyrol. 2007;78(2):257–64.
10. Wang S, Slovak V, Haynes BS. Kinetic studies of graphon and coal-char reaction with NO and O_2: direct nonlinear regression from TG curves. Fuel Process Technol. 2005;86(6): 651–60.
11. de Jong W, Di Nola G, Venneker BCH, Spliethoff H, Wójtowicz MA. TG-FTIR pyrolysis of coal and secondary biomass fuels: determination of pyrolysis kinetic parameters for main species and NO_x precursors. Fuel. 2007;86:2367–76.
12. Gui-qiu SU, Changlin CUI, Hongbo LU. Analysis of the smoke of pyrolysis and combustion of coal based on TG-FTIR method. Ind Boiler. 2004;2:23–6.
13. Zhu W, Zhou Y, Yang J, Chang A, Zhao H. Study on coal combustion features with thermal balance. Coal Sci Technol. 2004;32(3):8–11.
14. Khraisha YH. Flash pyrolysis of oil shales in a fluidized bed reactor. Energy Convers Manag. 2000;41:1729–39.
15. Ouyang Shan, Yeasmin H. A pressurized drop-tube furnace for coal reactivity studies. Rev Sci Instrum. 1998;69(8):3036–41.
16. Fletcher Thomas H. Time-resolved temperature measurements of individual coal particles during devolatilization. Combust Sci Technol. 1989;63:89–105.

17. Cho HC, Cho KW, Lee YK, Shin HD. Effect of coal type on gasification in pressurized drop tube furnace. Int J Energy Res. 2001;25(5):427–38.
18. Yu J, Harris D, Lucas J, Roberts D, Wu H, Wall T. Effect of pressure on char formation during pyrolysis of pulverized coal. Energy Fuels. 2004;18:1346–53.
19. Lu L, Sahajwalla V, Harris D. Characteristics of chars prepared from various pulverized coals at different temperatures using drop tube furnace. Energy Fuels. 2000;14:869–76.
20. Hindmarsh CJ, Thomas KM, Wang WX. A comparison of the pyrolysis of coal in wire-mesh and entrained-flow reactors. Fuel. 1995;74(8):1185–90.
21. Khraisha YH. Flash pyrolysis of oil shales in a fluidized bed reactor. Energy Convers Manag. 2000;41:1729–39.
22. Monson CR, Germane GJ. A high-pressure drop-tube facility for coal combustion studies. Energy Fuels. 1993;9(7):928–36.
23. Gale TK, Bartholomew CH, Fletcher TH. Effects of pyrolysis heating rate on intrinsic reactivities of coal chars. Energy Fuels. 1996;10:766–75.
24. Chen W-H. A simplified model of predicting coal reaction in a partial oxidation environment. Int Commun Heat Mass Transf. 2007;34:623–9.
25. Ulloa C, Borrego AG, Hello S, Gordon AL, et al. Char characterization and DTF assays as tools to predict burnout of coal blends in power plants. Fuel. 2005;84:247–57.
26. Kajitani S, Hara S, Matsuda H. Gasification rate analysis of coal with a pressurized drop tube furnace. Fuel. 2002;81:539–46.
27. Ballester J, García-Armingol T. Diagnostic techniques for the monitoring and control of practical flames. Prog Energy Combust Sci. 2010;36(4):375–411.
28. Dass C. Fundamentals of contemporary mass spectrometry. Hoboken: Wiley; 2007. p. 3–11.
29. Cen K. Boiler combustion experimental methods and measurement techniques. Beijing: China Power Press; 1987.
30. Li WH, Lou C, Sun YP, et al. Estimation of radiative properties and temperature distributions in coal-fired boiler furnaces by a portable image processing system. Exp Thermal Fluid Sci. 2011;35(2):416–21.
31. Zhou HC, Lou C, Cheng Q, et al. Experimental investigations on visualization of three-dimensional temperature distributions in a large-scale pulverized-coal-fired boiler furnace. Proc Combust Inst. 2005;30(1):1699–706.
32. Gong Y, Guo QH, Liang QF, et al. Three-dimensional temperature distribution of impinging flames in an opposed multiburner gasifier. Ind Eng Chem Res. 2012;51(22):7828–37.
33. Zhou B, Zhang J, Wang S. Reconstruction of flame temperature field with optical sectioning method. IET Image Proc. 2011;5(5):382–93.
34. Liu D, Yan JH, Wang F, et al. Experimental reconstructions of flame temperature distributions in laboratory-scale and large-scale pulverized-coal fired furnaces by inverse radiation analysis. Fuel. 2012;93(1):397–403.
35. Huang Q, Wang F, Yan J, Chi Y. Theoretical fast non-intrusive 3-D temperature distribution measurement within scattering medium from flame emission image analysis. Opt Commun. 2013;292:25–30.
36. Liu D, Huang QX, Wang F, et al. Simultaneous measurement of three-dimensional soot temperature and volume fraction fields in axisymmetric or asymmetric small unconfined flames with CCD cameras. J Heat Transfer. 2010;132:0612021–7.
37. Huang Q, Wang F, Yan J, Chi Yong. Simultaneous estimation of the 3-D soot temperature and volume fraction distributions in asymmetric flames using high-speed stereoscopic images. Appl Opt. 2012;51(15):2968–78.
38. Hossain MM, Lu G, Yan Y. Measurement of flame temperature distribution using optical tomographic and two-color pyrometric techniques, 2012.
39. Kashif M, Bonnety J, Guibert P, et al. Soot volume fraction fields in unsteady axis-symmetric flames by continuous laser extinction technique. Opt Express. 2012;20(27):28742–51.
40. Hossain MM, Lu G, Yan Y. Optical fiber imaging based tomographic reconstruction of burner flames. IEEE Trans Instrum Meas. 2012;61(5):1417–25.

41. Anikin NB, Suntz R, Bockhorn H. Tomographic reconstruction of 2D-OH*-chemiluminescence distributions in turbulent diffusion flames. Appl Phys B Lasers Opt. 2012;107 (3SI):591–602.
42. Chen YY, Zheng GG, Gu F, et al. Flow field's temperature partition reconstruction based on its phase distribution. Opt Commun. 2012;285(10–11):2501–5.
43. Ishino Y, Ohiwa N. Three-dimensional computerized tomographic reconstruction of instantaneous distribution of chemiluminescence of a turbulent premixed flame[J]. JSME Int J Ser B Fluids Therm Eng. 2005;48(1):34–40.
44. Lammel G, Schweizer S, Renaud P. MEMS infrared gas spectrometer based on a porous silicon tunable filter. Micro Electro Mech Syst. 2001;8:578–81.
45. Eaton AM, Smoot LD, Hill SC, et al. Components, formulations, solutions, evaluation, and application of comprehensive combustion models. Prog Energy Combust Sci. 1999;25 (4):387–436.
46. Fluent-Inc. Fluent 6.3. User's Guide, 2006.
47. Huang L, Li Z, Sun R, et al. Numerical study on the effect of the Over-Fire-Air to the air flow and coal combustion in a 670 t/h wall-fired boiler. Fuel Process Technol. 2006;87(4):363–71.
48. Vikhansky A, Bar-Ziv E, Chudnovsky B, et al. Measurements and numerical simulations for optimization of the combustion process in a utility boiler. Int J Energy Res. 2004;28(5): 391–401.
49. Zeng L, Li Z, Zhao G, et al. Numerical simulation of combustion characteristics and NO_x emissions in a 300 MWe utility boiler with different outer secondary-air vane angles. Energy Fuels. 2010;24(10):5349–58.
50. Vuthaluru HB, Vuthaluru R. Control of ash related problems in a large scale tangentially fired boiler using CFD modelling. Appl Energy. 2010;87(4):1418–26.
51. Díez LI, Cortés C, Pallarés J. Numerical investigation of NO_x emissions from a tangentially-fired utility boiler under conventional and overfire air operation. Fuel. 2008;87 (7):1259–69.
52. Dong C, Yang Y, Yang R, et al. Numerical modelling of the gasification based biomass co-firing in a 600 MW pulverized coal boiler. Appl Energy. 2010;87(9):2834–8.
53. Modlinski N. Computational modelling of a utility boiler tangentially-fired furnace retrofitted with swirl burners. Fuel Process Technol. 2010;91(11):1601–8.
54. Belosevic S, Sijercic M, Oka S, et al. Three-dimensional modelling of utility boiler pulverized coal tangentially fired furnace. Int J Heat Mass Transf. 2006;49(19):3371–8.
55. Asotani T, Yamashita T, Tominaga H, et al. Prediction of ignition behavior in a tangentially fired pulverized coal boiler using CFD. Fuel. 2008;87(4):482–90.
56. Belosevic S, Sijercic M, Tucakovic D, et al. A numerical study of a utility boiler tangentially-fired furnace under different operating conditions. Fuel. 2008;87(15):3331–8.
57. Fang Q, Musa AAB, Wei Y, et al. Numerical simulation of multifuel combustion in a 200 MW tangentially fired utility boiler. Energy Fuels. 2011;26(1):313–23.
58. Al-Abbas AH, Naser J, Dodds D. CFD modelling of air-fired and oxy-fuel combustion in a large-scale furnace at Loy Yang A brown coal power station. Fuel. 2012;102:646–65.
59. Tian ZF, Witt PJ, Schwarz MP, et al. Combustion of predried brown coal in a tangentially fired furnace under different operating conditions. Energy Fuels. 2012;26(2):1044–53.
60. Belosevic S, Beljanski V, Tomanovic I, et al. Numerical analysis of NO_x control by combustion modifications in pulverized coal utility boiler. Energy Fuels. 2012;26(1):425–42.
61. Choi CR, Kim CN. Numerical investigation on the flow, combustion and NO_x emission characteristics in a 500 MWe tangentially fired pulverized-coal boiler. Fuel. 2009;88 (9):1720–31.
62. Tian ZF, Witt PJ, Schwarz MP, et al. Numerical modelling of Victorian brown coal combustion in a tangentially fired furnace. Energy Fuels. 2010;24(9):4971–9.
63. Belosevic S, Sijercic M, Crnomarkovic N, et al. Numerical prediction of pulverized coal flame in utility boiler furnaces. Energy Fuels. 2009;23(11):5401–12.
64. Yan L, He B, Yao F, et al. Numerical simulation of a 600 MW utility boiler with different tangential arrangements of burners. Energy Fuels. 2012;26(9):5491–502.

65. Zhou H, Mo G, Si D, et al. Numerical simulation of the NO_x emissions in a 1000 MW tangentially fired pulverized-coal boiler: influence of the multi-group arrangement of the separated over fire air. Energy Fuels. 2011;25(5):2004–12.
66. Ren F, Li Z, Liu G, et al. Numerical simulation of flow and combustion characteristics in a 300 MWe down-fired boiler with different overfire air angles. Energy Fuels. 2011;25(4): 1457–64.
67. Fang Q, Wang H, Zhou H, et al. Improving the performance of a 300 MW down-fired pulverized-coal utility boiler by inclining downward the F-layer secondary air. Energy Fuels. 2010;24(9):4857–65.
68. Pallarés J, Gil A, Cortés C, et al. Numerical study of co-firing coal and *Cynara cardunculus* in a 350 MWe utility boiler. Fuel Process Technol. 2009;90(10):1207–13.
69. Agraniotis M, Stamatis D, Grammelis P, et al. Dry lignite cofiring in a greek utility boiler: experimental activities and numerical simulations. Energy Fuels. 2010;24(10):5464–73.
70. Taha TJ, Stam AF, Stam K, et al. CFD modelling of ash deposition for co-combustion of MBM with coal in a tangentially fired utility boiler. Fuel Process Technol. 2013;114:126–34.
71. Al-Abbas AH, Naser J, Hussein EK. Numerical simulation of brown coal combustion in a 550 MW tangentially-fired furnace under different operating conditions. Fuel. 2013;107: 688–98.
72. Liu ZP, Huang YL. Fuzzy model-based optimal dispatching for NO_x reduction in power plants. Int J Electr Power Energy Syst. 1998;20(3):169–76.
73. Kalogirou SA. Artificial intelligence for the modeling and control of combustion processes: a review. Prog Energy Combust Sci. 2003;29(6):515–66.

Chapter 4
Neural Network Modeling of Combustion Characteristics

Abstract In this chapter, artificial neural network (ANN) model is discussed with the basic idea and principle. It has been applied in many AI areas and industrial fields including pattern recognition and speech recognition. Then, two typical kinds of ANN algorithm (BPNN and GRNN) are introduced in this section. Through the comparative analysis of the principle and the conditions of the two methods of BPNN and GRNN, a new method combining the two methods is proposed.

Keywords Artificial neural network (ANN) model · Structural model of neuron
Back propagation neural network (BPNN) · Learning methods · General regression neural network (GRNN) · Comparison of BPNN method and GRNN method
Predicting coal ash fusion temperature · Generalization ability

4.1 Introduction

Neural network is composited by amount of the connecting neurons, which is the processing unit. Artificial neural network (ANN) model is proposed based on the modern neural research, it is a simplified and simulative model of biological neural networks [1, 2]. ANN has been successfully applied in pattern recognition [3], speech recognition, expert system et al., the basis idea and principle will be discussed in the next section.

4.1.1 Structural Model of Neuron

Neuron is the basis processing unit of neural work, it is usually regarded as a multi-input list output nonlinear model, as shown in Fig. 4.1.

u_i represents internal state of neuron i, θ_i is the threshold value, x_i is input, w_{ij} represents weight of connection of neuron x_i, s_i is the output control signal [4].

Neuron model is usually described as Eq. (4.1):

H. Zhou and K. Cen, *Combustion Optimization Based on Computational Intelligence*, Advanced Topics in Science and Technology in China,
https://doi.org/10.1007/978-981-10-7875-0_4

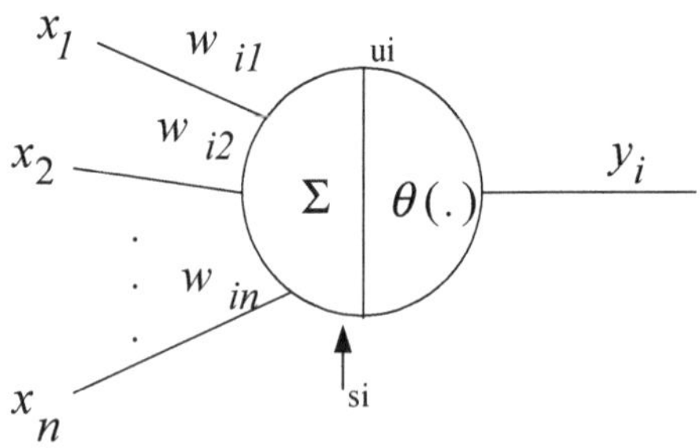

Fig. 4.1 Neuron model

$$\begin{cases} \tau \frac{du_i}{d_t} = -u_i(t) + \sum w_{ij}x_j(t) - \theta_i \\ y_i(t) = f[u_i(t)] \end{cases} \tag{4.1}$$

The output of neuron is expressed according to the different excitation function *f*.

(1) The threshold type.

$$f(u_i) = \begin{cases} 1, & u_i \geq 0 \\ 0, & u_i < 0 \end{cases} \tag{4.2}$$

(2) The piecewise linearity type.

$$f(u_i) = \begin{cases} 1, & u_i \geq u_2 \\ au_i + b, & u_1 < u_i < u_2 \\ 0, & u_i \leq u_1 \end{cases} \tag{4.3}$$

(3) S type.

$$f(u_i) = \frac{1}{1 + e^{-au_i}} \tag{4.4}$$

4.1.2 MP Model

MP model is proposed by Mc Culloch firstly, it has been involved to be the standard neural network. A stand MP model is shown as Fig. 4.2.

Fig. 4.2 MP model

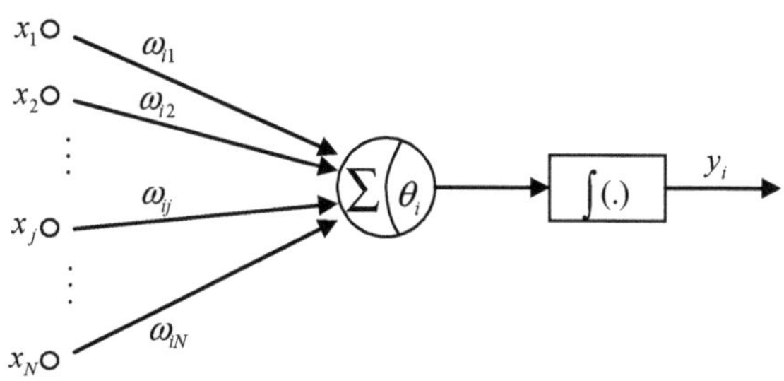

MP model is similar to neuron model, for the *i*th neuron in the network, it receives the signal from other neurons x_j (j = 1, …, N), the joint strength is represented as w_{ij}, means the weight of effect of neuron j to neuron i. We combine all the input according to a specific function, called net input. Net input exists many types of form, the simplest one is linear weighted summation:

$$I_i = \sum_j \omega_{ij} x_j \tag{4.5}$$

The mathematical expression of MP model is:

$$y_i = f\left(\sum_j \omega_{ij} x_j - \theta_i\right) \tag{4.6}$$

where θ_i is the threshold value of neuron i. The most commonly used excitation function is S type.

MP model doesn't give any learning methods to adjust the weight of connection in origin. Therefore, in the next section, some learning algorithm will be discussed.

4.2 Back Propagation Neural Network Method

4.2.1 BPNN Algorithm

Back propagation neural network (BPNN) is a kind of multilayer feed-forward neural network, it is put forward by Rumelhart in 1985 to gain better classify ability. BPNN adopts supervised learning and gradient descent algorithm in modeling process. Topological structure with only one hidden layer of BPNN is shown as Fig. 4.3.

BPNN is composed by two parts: forward-propagating and back propagation. During forward propagation, input information reaches out layer after handing in hidden layer, every neural layer is only influenced by the upper layer. Weight of neuron connection stay unchanged during forward propagation. If the output

Fig. 4.3 Topological structure of BPNN

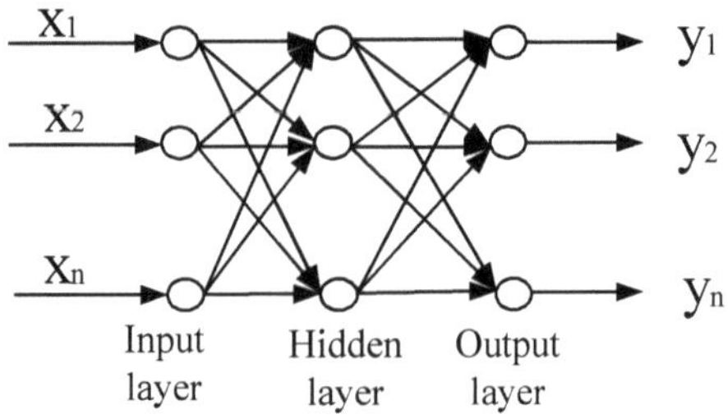

disagrees with the desired value, the error counter propagates. During the back propagation, the error signal is the difference between output and desired value, error information transmits from the output layer. In this process, network weight is modified according to the feedback to hold output to the desired value.

4.2.2 Learning Methods

Two kinds signal of BPNN is shown as Fig. 4.4.

Operating signal (solid line). It is the signal that generated after the input propagates to the output layer, operating signal is the function of weights of input.

Error signal (imaginary line). Error is the difference between output of the neural network and the desired value, error signal propagates backwards from the output layer.

Given $y_i(n)$ as the output of *j*th unit at the *n* iteration, the error of this unit is:

$$e_j(n) = d_j(n) - y_j(n) \tag{4.7}$$

Defining the square error of the *j* unit is $\frac{1}{2}e_j^2(n)$, therefor the total error of output is:

$$E(n) = \frac{1}{2}\sum e_j^2(n) \tag{4.8}$$

Fig. 4.4 Two kinds of signal of BPNN

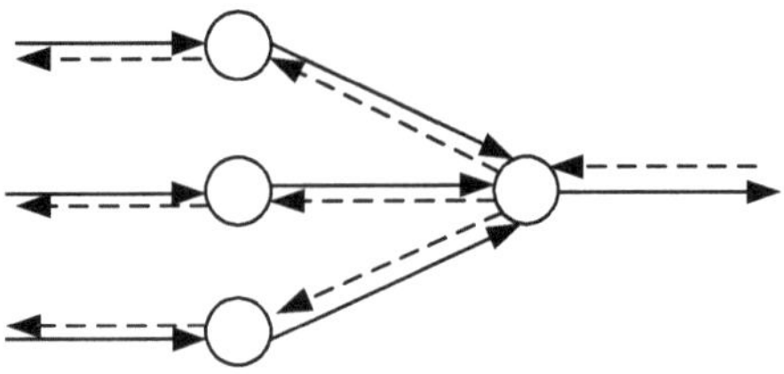

Assuming number of samples in train set is N, the average square error is:

$$E(AV) = \frac{1}{N}\sum_{1}^{N} E(n) \tag{4.9}$$

$E(AV)$ is the objective function of learning method, learning purpose aims to minimize $E(AV)$, $E(AV)$ is the function of network weights, threshold value and input.

Setting $v_j(n) = \sum_{i=0}^{p} w_{ji}(n)y_i(n)$, p is the number of input added into unit j, so we can get $y_j(n) = f_j(v_j(n))$, where f_i is the excitation function. Therefore, the correction of w_{ji} is:

$$\Delta w_{ji} = -\eta \frac{\partial E(n)}{\partial w_{ji}(n)} = \eta \delta_j(n) y_i(n) \tag{4.10}$$

where negative indication means the direction of gradient descent, $\delta_j(n) = e_j(n)f'(v_j(n))$ called local gradients. Local gradients are divided into two kinds:

(1) unit j is the output;

$$\delta_j(n) = (d(n) - y_j(n))f'(v_j(n)) \tag{4.11}$$

(2) unit j is the hidden layer.

$$\delta_j(n) = f_j'(v_j(n)) \sum_k \delta_k(n) w_{kj}(n) \tag{4.12}$$

In the actual application, training set should be imported, every output of all training samples is called training cycle, learning method is implemented by repeating the training cycle until the objective function reaches the target.

4.3 General Regression Neural Network Method

General regression neural network (GRNN) is one of the radial basis function (RBF), proposed by Specht in 1991, aims to solve functional approximation problem [5, 6].

4.3.1 GRNN Algorithm

Assuming during a process, the input is $X = [x_1, x_2, \ldots, x_p]$, p is the dimension, output Y is a scalar, decided by X. Given a samples set:$[(X_i, Y_i)]$, $i = 1, 2, \ldots, n$, Y_i is the observed value. GRNN aims to notarize the relationship between X and Y. In general regression, Y is considered as the regression about X:

$$\hat{Y}(X) = E[y|X] = \frac{\int_{-\infty}^{+\infty} dyy \cdot f(X, y)}{\int_{-\infty}^{+\infty} dy \cdot f(X, y)} \tag{4.13}$$

where $f(X, y)$ is joint probability distribution function of independent variable X and dependent variable Y, $\hat{Y}(X)$ is the expectancy value of certain X. For any arbitrary X, output Y can be calculated according to (4.13).

In generally, $f(X, y)$ is unknown. Here we define some parameters.

$$\text{Euclidean distance:} \quad D(X_i, X_j) = \sqrt{(X_i, X_j)^{\mathrm{T}}(X_i, X_j)} \tag{4.14}$$

$$\text{Central function:} \quad W(X) = \exp\left(-\frac{X^2}{2}\right) \tag{4.15}$$

For a given set $[(X_i, Y_i)]$, $i = 1, 2, \ldots, n$, joint probability distribution function is defined as:

$$\hat{Y}(X) = \frac{\sum_i^n Y_i \cdot W\left(\frac{D(X,X_i)}{\sigma}\right)}{\sum_i^n W\left(\frac{D(X,X_i)}{\sigma}\right)} \tag{4.16}$$

where

$$\begin{aligned} \lim_{n\to\infty} \sigma(n) &= 0 \\ \lim_{n\to\infty} n\sigma^p(n) &= \infty \end{aligned} \tag{4.17}$$

According to (4.16), as long as given a set $[(X_i, Y_i)]$, any new Y corresponding of X can be determined.

4.3.2 GRNN Structure

GRNN is formed by four layers: input layer, pattern layer, summation layer and output layer, as shown in Fig. 4.5.

Fig. 4.5 Structure of GRNN

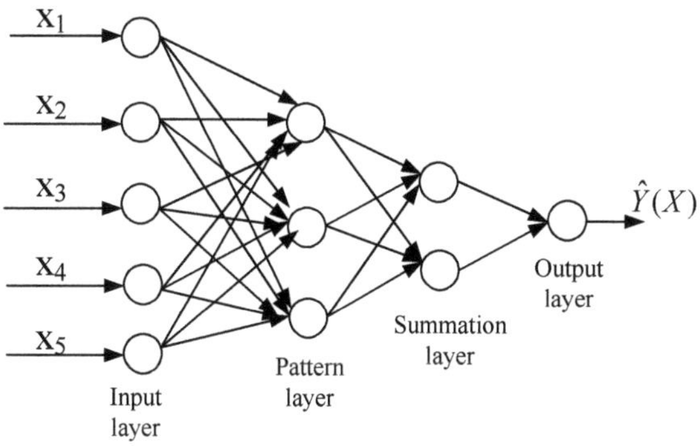

Input layer accepts input signal only, it distributes variable and transforms information to the pattern layer. When new variable is imported into the network, Pattern layer accepts the distributed input variables, calculates deviation of the distance between new variable and observed value. Summation layer acquires net weights and dot product of pattern layer output. Output layer exports the estimated value $\hat{Y}(X)$.

4.4 Comparison of BPNN Method and GRNN Method

4.4.1 GRNN Advantages

Compared with BPNN, GRNN has many advantages:

(1) Net structure and weight are totally certain.
(2) Only one model parameter needs training.
(3) GRNN accuracy is adjustable.

4.4.2 Comparison on Example

The comparison is carried out on the predicting coal ash fusion temperature, the relative error of BPNN and GRNN is shown in Fig. 4.6.

GRNN performs better in generalization ability, BPNN also consumes more time in training process with reduplicative modified net weight. Another matter should be noticed is training cases in BPNN model must be more than 10 times of connection weight. In this comparison, train set size is 30, which is much more less than the needed 900 cases. GRNN only requires 1% of BPNN in train set size to obtain the same accuracy. So GRNN is more suitable for small samples model.

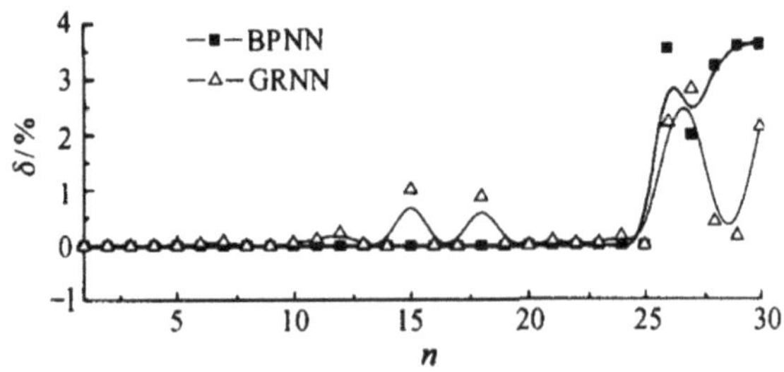

Fig. 4.6 Comparison between predicting performance pf GRNN and BPNN

4.5 Summary

In this section, ANN is discussed. ANN is a simplified and simulative model of biological neural networks, it has achieved successful application in many AI area and industrial fields. The basis idea and structure of ANN is introduced.

Two main kinds of ANN algorithm are put forward: BPNN and GRNN. BPNN is a widely used algorithm is nonlinear computation, GRNN adopt some variant in training process to gain fast modeling speed and predictive accuracy. Comparison between BPNN and GRNN shows GRNN is superior to BPNN.

References

1. Jayas DS, Paliwal J, Visen NS. Review paper (AE—automation and emerging technologies): multi-layer neural networks for image analysis of agricultural products. J Agric Eng Res. 2000;77(2):119–28.
2. Yegnanarayana B. Artificial neural networks. PHI Learning Pvt. Ltd., 2004.
3. Specht DF. A general regression neural network. IEEE Trans Neural Netw. 1991;2(6):568–76.
4. Huang SH, Zhang HC. Artificial neural networks in manufacturing: concepts, applications, and perspectives. IEEE Trans Compon Packag Manuf Technol Part A. 1994;17(2):212–28.
5. Xi-guang S, Li-gang Z, Hao Z. Combining general regression neural network and genetic algorithm to optimize ash fusion temperature. J Zhejiang Univ (Eng Sci). 2005;39(8):1189–92.
6. Zhou H, Zheng LG, Fan JR, et al. Application of general regression neural network in prediction of coal ash fusion temperature. Zhejiang Daxue Xuebao (Gongxue Ban)/J Zhejiang Univ (Eng Sci). 2004;38(11):1479–82.

Chapter 5
Classification of the Combustion Characteristics based on Support Vector Machine Modeling

Abstract In this chapter, support vector machine (SVM) based on the structural risk minimization principle is exposed. As a vital computational method with outstanding generalization ability, it has been widely used in the field of machine learning and data mining. Support vector classification (SVC) and support vector regression (SVR) are the main parts of SVM [1]. The principle of them is introduced in this section. In the field of coal-fired power generation, there are lots of applications of SVC and SVR in the simulation. They are conducted to the modeling of coal identification, prediction of NO_x emission, prediction of unburned carbon in fly ash and others.

Keywords Support vector machine · Machine learning and data mining
Structural risk minimization principle · Support vector classification (SVC)
Support vector regression (SVR) · Coal identification · Prediction of ash fusion temperature · Prediction of unburned carbon · Prediction of NO_x emission

5.1 The Introduction of Support Vector Machine

Support vector machine is a very popular method in the field of machine learning and data mining [2–5]. It is based on the structural risk minimization principle, and considers the generalization ability and experience risk. Therefore, support vector machine has better generalization ability than that of neural network and other machine learning methods, and is considered to be very suitable for processing small and multidimensional samples. It is universal, robust, and effective to use for classification and regression analysis. Its performance is very good in many challenging applications. The support vector machine is one of the best modeling tools for the data which is lack of specific model.

The basic support vector machine constructs a hyperplane in space, which can be used for classification. The optimal hyperplane is here defined as the linear decision function with maximal margin between the vectors of the two classes, which is shown in Fig. 5.1. It was observed that to construct such optimal hyperplanes,

H. Zhou and K. Cen, *Combustion Optimization Based on Computational Intelligence*, Advanced Topics in Science and Technology in China,
https://doi.org/10.1007/978-981-10-7875-0_5

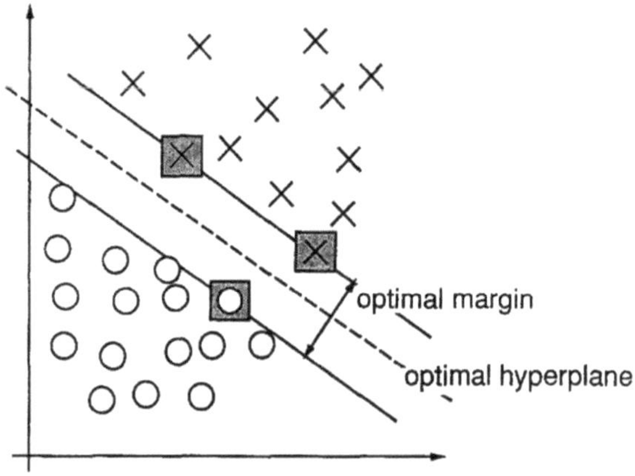

Fig. 5.1 An example of support vector classification in a two-dimensional space

one only has to take into account a small amount of the training data, the so-called support vectors, which determine the margin and so, it is called support vector classification (SVC). The original optimal hyperplane algorithm was proposed in 1963. It was a linear classifier. Given a set of training examples, each marked as belonging to one of two categories, an SVC training algorithm builds a model that assigns new examples into one category or the other. An SVC model is a representation of the examples as points in space, so that the examples of the separate categories are divided by a clear margin that is as wide as possible. New examples are then predicted to belong to a category based on which side of the margin they fall on.

In addition to performing linear classification, SVC can efficiently perform a nonlinear classification by using kernel trick, implicitly mapping their inputs into high-dimensional feature spaces. Whereas the original problem may be stated in a finite dimensional space, it often happens that the sets to discriminate are not linearly separable in that space. In 1992, a method to create nonlinear classifiers by applying the kernel trick to maximum margin hyperplanes was proposed. The resulting algorithm is formally similar, except that every dot product is replaced by a nonlinear kernel function. This allows the algorithm to fit the maximum margin hyperplane in the transformed feature space. The transformation may be nonlinear, though the classifier is a hyperplane in the high-dimensional feature space, it may be nonlinear in the original input space.

A version of SVM for regression was proposed in 1996. This method is called support vector regression (SVR). The model produced by SVC (as described above) depends only on a subset of the training data, because the cost function for building the model does not care about training points that lie beyond the margin. Analogously, the model produced by SVR depends only on a subset of the training data, because the cost function for building the model ignores any training data close to the model prediction.

5.2 The Principle of Support Vector Machine

5.2.1 *Support Vector Classification*

Consider training vectors $\boldsymbol{x}_i \in \mathrm{R}^m$, $i = 1, 2, \ldots, n$, in two classes, and an indicator vector $\boldsymbol{y} \in \mathrm{R}^n$ such that $y_i \in \{1, -1\}$. The aim of support vector classification (SVC) is to construct a linear decision boundary which separates training vectors $\boldsymbol{x}_i$ to two classes with maximal margin.

SVC mainly solves the following optimization problem:

$$\min_{w,b,\xi} \left\{ \frac{1}{2} w^T w + C \sum_{i=1}^{n} \xi_i \right\}$$

$$\text{subject to} \quad y_i(w^T \Phi(\boldsymbol{x}_i) + b) \geq 1 - \xi_i$$
$$\xi_i \geq 0, \; i = 1, 2, \ldots, n,$$

where $\Phi(\boldsymbol{x}_i)$ maps $\boldsymbol{x}_i$ into a higher dimensional space. C is the penalty parameter, and ξ_i is the slack variable.

The optimization problem above is often solved by its dual problem

$$\min_{\boldsymbol{a}} \left\{ \frac{1}{2} \boldsymbol{a}^T Q \boldsymbol{a} - \boldsymbol{e}^T \boldsymbol{a} \right\}$$

$$\text{subject to} \quad \boldsymbol{a}^T \boldsymbol{y} = 0$$
$$0 \leq \boldsymbol{a} \leq C\boldsymbol{e},$$

where $\boldsymbol{a} = [a_1\, a_2, \ldots, a_n]$ and a_i is the Lagrange multiplier, $\boldsymbol{e} = [1\, 1, \ldots, 1]^{\mathrm{T}}$ is the vector of all ones, Q is an n by n positive matrix, $Q_{ij} \equiv y_i y_j K(x_i, x_j)$, and $K(x_i, x_j) \equiv \Phi(x_i)^{\mathrm{T}} \Phi(x_j)$ is the kernel function.

After solving the dual problem, the original optimization problem is also solved and the linear decision boundary can be written as follows:

$$\mathrm{sgn}\left(\sum_{i=1}^{n} y_i a_i K(\boldsymbol{x}_i, \boldsymbol{x}) + b \right)$$

For multi-class classification problem, "one-against-one" approach can be implemented. If k is the number of classes, then $k(k-1)/2$ classifiers are constructed and each one trains data from two classes. The classification is determined by a voting strategy.

5.2.2 Support Vector Regression

Given a set of training points, $\{(x_1, z_1), (x_2, z_2), \ldots, (x_n, z_n)\}$, where $\boldsymbol{x_i} \in R^m$ and is a z_i is the target output. The aim of support vector regression (SVR) is to construct a linear decision boundary which fits training points with minimal margin.

SVR mainly solves the following optimization problem:

$$\min_{w,b,\xi,\xi^*} \left\{ \frac{1}{2} \boldsymbol{w^T w} + C \sum_{i=1}^{n} \xi_i + C \sum_{i=1}^{n} \xi_i^* \right\}$$

$$\begin{aligned} \text{subject to} \quad & \boldsymbol{w^T} \Phi(\boldsymbol{x_i}) + b - z_i \geq \varepsilon + \xi_i \\ & z_i - \boldsymbol{w^T} \Phi(\boldsymbol{x_i}) - b \leq \varepsilon + \xi_i^*, \\ & \xi_i, \xi_i^* \geq 0, \ i = 1, 2, \ldots, n \end{aligned}$$

where $\Phi(\boldsymbol{x_i})$ maps $\boldsymbol{x_i}$ into a higher dimensional space. C is the penalty parameter, ξ_i and ξ_i^* are the slack variables, and ε is the prediction error.

The optimization problem above is also often solved by its dual problem:

$$\min_{a,a^*} \left\{ \frac{1}{2} (\boldsymbol{a} - \boldsymbol{a}^*)^T \boldsymbol{Q} (\boldsymbol{a} - \boldsymbol{a}^*) + \varepsilon \sum_{i=1}^{n} (a_i - a_i^*) + \sum_{i=1}^{n} z_i (a_i - a_i^*) \right\}$$

$$\begin{aligned} \text{subject to} \quad & \boldsymbol{e}^T (\boldsymbol{a} - \boldsymbol{a}^*) = 0 \\ & 0 \leq \boldsymbol{a}, \boldsymbol{a}^* \leq C\boldsymbol{e}, \end{aligned}$$

where $\boldsymbol{a} = [a_1 a_2, \ldots, a_n], \boldsymbol{a}^* = [a_1^* a_2^*, \ldots, a^*]$, a_i and a_i^* are the Lagrange multipliers, $e = [1\,1, \ldots, 1]^T$ is the vector of all ones, Q is an n by n positive matrix, $Q_{ij} \equiv y_i y_j K(x_i, x_j)$, and $K(x_i, x_j) \equiv \Phi(x_i)^T \Phi(x_j)$ is the kernel function.

After solving the dual problem, the original optimization problem is also solved and the approximate function can be written as follows:

$$\sum_{i=1}^{n} (a_i^* - a_i) K(\boldsymbol{x_i}, \boldsymbol{x}) + b$$

5.2.3 Kernel Function

Often, the original samples are usually not linear separable. But the linear decision boundary can still be found in a higher dimensional space, where the samples are mapped to. It is difficult to find the mapping. But fortunately, the dot product in a higher dimensional space can be calculated by kernel function.

The kernel function mainly includes linear kernel function, polynomial kernel function, and radial basis kernel function. They are defined as follows:

$$\text{Linear:}\quad K(\boldsymbol{x}_i^{\cdot}, \boldsymbol{x}_j^{\cdot}) = <\boldsymbol{x}_i^{\cdot} \cdot \boldsymbol{x}_j^{\cdot}>$$

$$\text{Polynomial:}\quad K(\boldsymbol{x}_i^{\cdot}, \boldsymbol{x}_j^{\cdot}) = (\gamma <\boldsymbol{x}_i^{\cdot} \cdot \boldsymbol{x}_j^{\cdot}> + r)^d$$

$$\text{Radial basis:}\quad K(\boldsymbol{x}_i^{\cdot}, \boldsymbol{x}_j^{\cdot}) = \exp(-g \parallel \boldsymbol{x}_i^{\cdot} - \boldsymbol{x}_j^{\cdot} \parallel^2),$$

where γ, r, d and g are kernel parameters.

Radial basis kernel function can achieve the nonlinear mapping and has less numerical difficulties. It is the most popular kernel function for use.

5.3 The Application of Support Vector Machine

5.3.1 Coal Identification

In power stations, many types of coals are available in stocks and the type of coal being fired is often unknown. This makes the application of combustion optimization be seriously restricted. Traditional coal analysis method is expensive and time-consuming. Therefore, online coal identification appears.

The online coal identification system is based on SVC and flame monitoring system (detail in Chap. 7) is often used for the data acquisition of coal identification. The following is an example of the application of coal identification system. The coal identification system, which is based on SVC, is tested on test combustion rig. The test combustion rig (Fig. 5.2) consisted of a vertical furnace, coal feeder, and swirl burner. The lengths of the vertical furnace and the inner diameter were 3950 and 350 mm, respectively. Refractory material was installed in the furnace to prevent the heat loss. The feed rate of the coal feeder could be varied from 10 to 45 kg/h by adjusting the rotational speed of the electromotor. The temperature of preheated primary air and swirl secondary air was about 40 °C and the air flow rate could be adjusted by valves.

Four types of coals were selected for collecting the flame futures and establishing the model for prediction. 18 flame features were extracted from flame images and light intensity signals. They are listed in Table 5.1. The flame features were obtained in variable operating conditions for every coal type. The operating conditions are shown in Table 5.2.

A total of 384 samples were obtained. They were divided into two groups randomly for 10 times. One group (288 samples) was used for training and the other group (96 samples) was used for testing. The ratio of the training group to the testing group was 3:1. After training, the testing group was used to assess the prediction accuracy. The prediction accuracies were listed in Table 5.3. The averaged prediction accuracy was 99.1% which meant that the coal identification was achieved successfully.

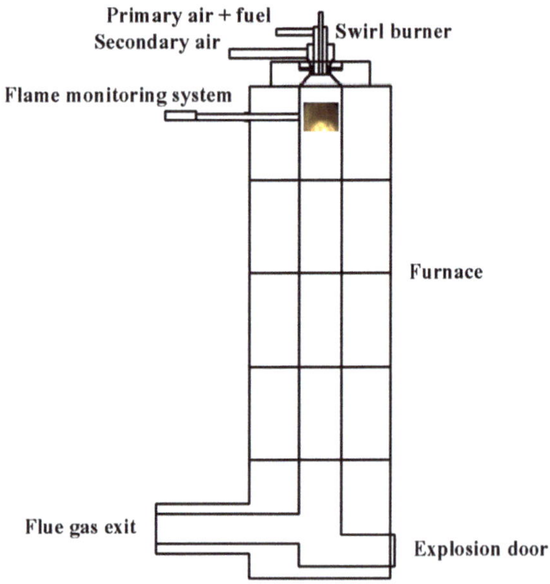

Fig. 5.2 Schematic diagram of pulverized coal combustion furnace and flame monitoring system

Table 5.1 Flame features

No.	Flame feature
1	Mean value of *R* channel
2	Mean value of *G* channel
3	Mean value of *B* channel
4	Standard deviation of *R* channel
5	Standard deviation of *G* channel
6	Standard deviation of *B* channel
7	Mean value of VI signal
8	Mean value of IR signal
9	Standard deviation of VI signal
10	Standard deviation of IR signal
11	Skewness of VI signal
12	Skewness of IR signal
13	Kurtosis of VI signal
14	Kurtosis of IR signal
15	Fluctuation ratio of VI signal
16	Fluctuation ratio of IR signal
17	Flicker of VI signal
18	Flicker of IR signal

Table 5.2 Experimental cases

Case	Coal feed rate (kg/h)	Primary air flow rate (kg/h)	Secondary air flow rate (kg/h)
1	25	54.2	157.7
2	25	59.0	171.8
3	25	64.5	182.7
4	25	59.0	149.9
5	25	64.5	165.5
6	25	70.1	174.9
7	30	64.5	189.0
8	30	69.3	203.0
9	30	74.9	220.2
10	30	69.3	182.7
11	30	75.7	198.3
12	30	81.3	213.9

Table 5.3 Prediction accuracy

No.	Accuracy (%)	No.	Accuracy (%)
1	100	6	98.96
2	98.96	7	100
3	100	8	96.88
4	98.96	9	98.96
5	97.92	10	100

5.3.2 The Prediction of Ash Fusion Temperature

Ash fusion temperature has a significant impact on slagging characteristics and thermal efficiency of the boiler. It is an important characteristic of the coal. Ash fusion temperature is influenced by ash composition, but there is not a clear and precision method to calculate the ash fusion temperature by ash composition. So, constructing the relationship model of ash fusion temperature and ash composition by SVR to predict ash fusion temperature is a viable approach [6, 7].

The following is an example of the application of ash fusion temperature prediction. Ash fusion temperature and ash compositions (SiO_2, Al_2O_3, Fe_2O_3, CaO, MgO, TiO_2, Na_2O, and K_2O) of 12 coal samples were measured. The results are shown in Table 5.4.

The model to predict ash fusion temperature was established by seven inputs and one output, SiO_2, Al_2O_3, Fe_2O_3, CaO, MgO, TiO_2, and Na_2O+K_2O as inputs and ash fusion temperature as output. Na_2O and K_2O have the similar influence on ash fusion temperature and their mass fraction was very small, so the mass fractions of Na_2O and K_2O were combined as one input.

Samples 3–12 were selected as training data, samples 1 and 3–12 were selected as testing data. The testing data both contains the trained data and the untrained

Table 5.4 Ash fusion temperature and ash compositions

No.	T_{st}/°C	F_2O_3/%	MgO/%	CaO/%	TiO_2/%	Al_2O_3/%	SiO_2/%	$K_2O + Na_2O$/%
1	1426	7.66	0.01	8.65	1.44	33.83	40.96	0.52
2	>1500	4.21	0.01	2.43	2.14	35.45	43.09	0.28
3	>1500	4.14	0.01	3.62	1.93	36.9	42.28	0.22
4	1157	10.08	1.23	15.49	0.95	12.57	41.54	1.08
5	1412	12.72	0.13	1.65	1.2	23.75	53.11	1.06
6	1210	8.16	2.19	17.84	1.26	13.72	39.42	0.67
7	1486	12.72	0.13	4.97	1.2	23.75	51.16	1.23
8	>1500	4.14	0.01	3	1.93	36.9	61.81	0.37
9	1312	4.21	0.01	6.25	2.14	35.45	53.18	0.68
10	>1500	10.08	1.23	2.97	0.95	12.57	46.73	0.15
11	1381	7.66	0.01	2.85	1.44	33.83	52.81	0.83
12	>1500	8.16	2.19	1.34	1.26	13.72	54.98	1.06

Table 5.5 Parameter optimization result

	g	*C*	Mean squared error
Value	>1500	8.16	2.19

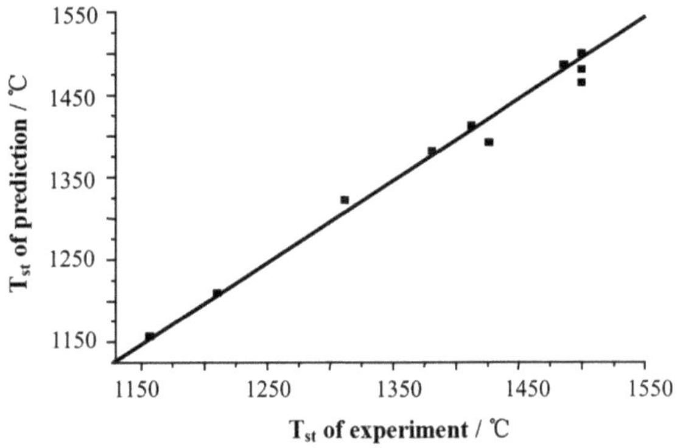

Fig. 5.3 The prediction result of ash fusion temperature

data, a better balance between empirical risk and generalization ability could achieve. After parameter optimization (the result of parameter optimization is listed in Table 5.5), the model was constructed to predict the ash fusion temperature. The result was shown in Fig. 5.3. The maximum relative error and mean relative error were 2.4 and 0.57%, respectively, which indicates that the model is precise for predicting the ash fusion temperature.

For predicting the ash fusion temperature of blending coals, 10 of 12 coal samples were chosen to mix with different ratios. The ash fusion temperature and ash compositions of blended coals are shown in Table 5.6.

Table 5.6 The parent coals and ratios of blending coals

Blending coals	11–12	9–11	2–3	5–6	1–7
Ratio	5:5 4:6 6:4	5:5	5:5 4:6 6:4	2:8 3:7 4:6 5:5 6:4 7:3 8:2	5:5 4:6 6:4

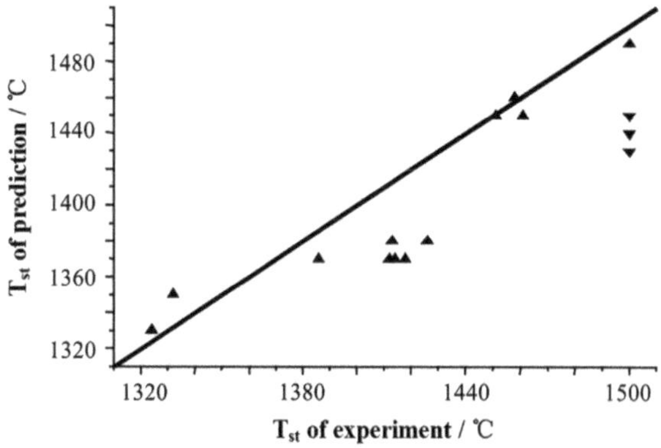

Fig. 5.4 The prediction result of ash fusion temperature of blending coals

After constructing the model, the model was used for predicting the ash fusion temperature of blending coals. The prediction result was shown in Fig. 5.4. The maximum relative error and mean relative error were 4.67 and 1.94%, respectively, which indicates that the model is also precise for predicting the ash fusion temperature of blending coals.

5.3.3 The Prediction of Unburned Carbon in Fly Ash

High levels of unburned carbon resulting from inefficient coal combustion lead to major environmental and economic consequences. Reductions of unburned carbon in ash are helpful to improve operational efficiency, and minimize solid waste for pulverized coal-fired boilers. Traditional measurement of unburned carbon in fly ash is expensive, therefore, SVR is introduced to predict the unburned carbon.

The following is an example of the application of unburned carbon prediction. The experiments was carried out at a 300 MW, tangentially fired dry bottom boiler with a furnace of 14.022 × 12.33 m^2 section and height of 48.5 m. The tilting fuel and combustion air nozzles including five primary air nozzles and six secondary air nozzles and one over fire air (OFA) nozzle are located in each corner of the furnace. All nozzles can be tilted in a vertical direction over ± 20° from the horizontal axis, both upward and downward to adjust the reheated steam temperature for the varying fouling conditions of the furnace. Four levels of the fuel burners are put into operation under the rated load. The medium-speed coal pulverizers are employed to supply the coal–air mixture to the burners on the corresponding levels. The dimensions of the furnace and the arrangement of the burners are illustrated in Fig. 5.5.

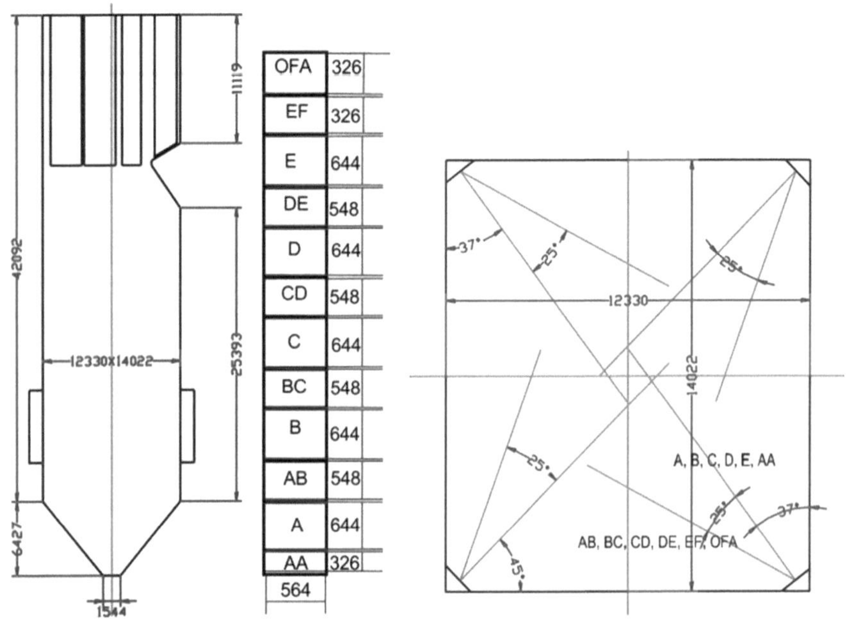

Fig. 5.5 The sketch of the investigated utility boiler

A total of 21 tests were performed on this boiler. The operating parameters of the tests are listed in Tables 5.7 and 5.8. Six coals are burned in the tests, and proximate analyses of the coals are also listed in Table 5.8. The damper position indicated in Tables 5.7 and 5.8 is the percentage opening of a butterfly valve across the area of the given inlet port.

During the experiments, coal fineness was kept constant, and the measured unburned carbon in fly ash is summarized in Table 5.9. The model to predict unburned carbon in fly ash was established by 21 inputs (Tables 5.7 and 5.8) and one output (Table 5.9).

For the prediction on single condition, conditions 1–10 and 12–21 were selected as training data to predict condition 11. For parameter optimization, first set *C* as default value to optimize *g*, then the best *g* was used to optimize *C*. The effects of *g* and *C* on predicting performance are shown in Figs. 5.6 and 5.7. The final parameter and capability of the model are listed in Table 5.10. The relative error is only 0.196%.

For the prediction on two conditions, conditions 2–15 and 17–21 were selected as training data to predict condition 1 and 16. After parameter optimization, the final parameter and capability of the model are listed in Table 5.11. The maximum relative error is 0.196%.

Table 5.7 The boiler operating conditions

Case	Load (MW)	The secondary air burner damper position (% open)					The feeding rate of the coal feeder (t/h)				
		AB	BC	CD	DE	EF	A	B	C	D	E
1	294.1	100	100	70	70	100	0	27	27	25	26
2	295.4	100	100	100	70	60	0	27	27	25	26
3	294.3	100	100	100	100	100	0	27	27	25	26
4	296.1	100	100	100	100	100	0	26.4	25.8	25.8	25.8
5	294	100	100	100	100	100	0	26	27	26	25
6	294.8	100	100	100	100	100	0	26.5	26	25.9	25.5
7	295.4	24	100	100	100	100	0	25	26	26	26
8	291.6	44	100	100	100	100	0	25	26	26	26
9	292.8	45	100	100	100	100	0	25	26	26	26
10	296	95	100	100	100	4	27	27	26	26	0
11	299.5	100	66	42	100	100	26	27	0	26	25
12	295.6	83	71	47	86	78	26	27	0	27	25
13	298.6	100	100	0	100	100	28	25	0	27	27
14	213.7	70	70	70	70	70	19	0	18	19	19
15	216.1	70	70	70	70	70	19	0	19	19	20
16	216.8	70	70	70	70	70	19	0	19	18	19
17	215.3	70	70	70	70	70	19	18	20	20	0
18	251.3	100	100	100	94	40	24	19	23	22	0
19	251	100	100	100	100	30	24	20	23	22	0
20	250.9	100	100	100	100	30	24	20	24	21	0
21	257.6	65	99	99	99	99	0	21	25	23	22

5.3.4 The Prediction of NO_x Emission

Coal combustion in modern power plant boilers generating most of global electricity emanates a large amount of NO_x to the ambient surroundings, subsequently results in severe hazards. Therefore, the control of NO_x is one of the most important problems to be solved in the operation of a coal-fired utility boiler [8–10].

However, the relationship between NO_x emissions characteristics and operating conditions is very complicated. It is very difficult to estimate NO_x emission by operating condition and find the best condition to reduce the NO_x emission. Therefore, SVR can be used for the modeling of NO_x emission and finding the best condition by global search tools.

The following is an example of the prediction of NO_x emission. The experiments have been performed in a dual-furnace tangentially fired full-scale boilers. It is a 300 MWe power tangentially fired boiler and is with a large furnace of 14.08 m 11.858 m cross section and 53 m high. Six secondary air ports and four primary air ports are arranged on each corner of the furnace. Coal is pulverized by four

Table 5.8 The boiler operating conditions

Case	O_2 (vol. %)	Damper position (% open)		Pressure drop (kPa)	The coal properties				Burner tilt (%)	Primary air pressure (kPa)
		Fuel air	OFA		V_{ar} (%)	A_{ar} (%)	M_{ar} (%)	$Q_{net,ar}$ (kJ/kg)		
1	3.54	100	85	0.71	28.92	11.552	8.1	25,350	50	7.8
2	3.47	100	83	0.75	28.92	11.552	8.1	25,350	50	7.52
3	3.05	80	0	0.796	28.26	17.019	7.87	23,364	100	7.58
4	3	79	84	0.656	28.26	17.019	7.87	23,364	100	7.84
5	2.88	20	84	0.795	28.26	17.019	7.87	23,364	100	7.82
6	3.14	47	83	0.74	29.32	13.076	11.57	22,222	20	7.69
7	2.09	81	87	0.555	29.94	11.522	11.6	23,820	100	7.45
8	3.67	81	82	0.78	29.94	11.522	11.6	23,820	100	7.25
9	4.02	83	82	0.82	29.94	11.522	11.6	23,820	100	7.5
10	3.4	50	88	0.572	27.17	12.798	11.1	22,636	20	7.52
11	3.14	50	85	0.67	27.17	12.798	11.1	22,636	20	7.17
12	3.09	20	87	0.765	27.17	12.798	11.1	22,636	20	7.6
13	2.35	80	90	0.565	27.17	12.798	11.1	22,636	50	7.6
14	4.27	38	34	0.56	26.225	17.436	11.7	21,810	100	6.9
15	3.68	38	33	0.43	26.225	17.436	11.7	21,810	100	6.8
16	5.05	38	31	0.79	26.225	17.436	11.7	21,810	100	7.58
17	4.45	38	33	0.5	26.225	17.436	11.7	21,810	100	7.1
18	3.89	85	60	0.5	26.225	17.436	11.7	21,810	50	8.3
19	3.08	85	58	0.4	26.225	17.436	11.7	21,810	50	8.2
20	4.8	85	57	0.67	26.225	17.436	11.7	21,810	50	8.2
21	4.07	85	58	0.7	26.225	17.436	11.7	21,810	100	7.5

Table 5.9 The unburned carbon content under various operating conditions

No.	1	2	3	4	5	6	7	8	9	10	11
Unburned carbon (%)	4.02	3.28	8.18	7.4	4.56	5.49	5.46	4.2	2.93	4.62	3.06
No.	12	13	14	15	16	17	18	19	20	21	
Unburned carbon (%)	2.76	2.35	1.94	2.25	1.64	1.64	2.34	2.9	1.52	3.88	

medium-speed pulverizers with rotating classifiers. The burner arrangement is shown in Fig. 5.8.

To obtain a wide range of operating conditions, a total of 670 tests have been carried out at this boiler to investigate the effects of the boiler load, primary air (PA), and secondary air (SA) distribution on NO_x emissions characteristics behavior. During the experiments, coal quality was kept constant, and NO_x concentrations were continuously monitored in the boiler outlet prior to the air heater by using

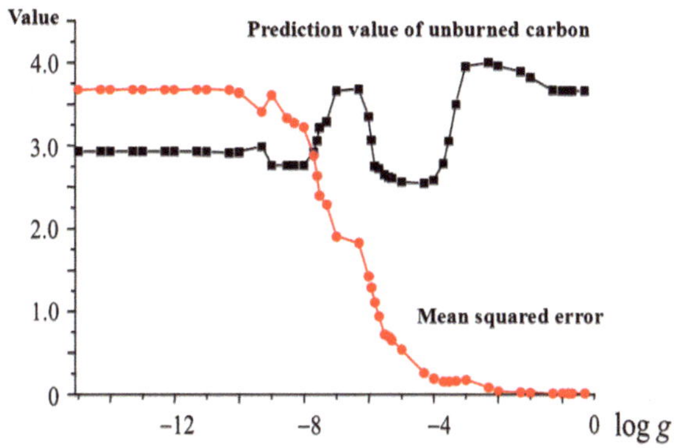

Fig. 5.6 The effects of *g* on predicting performance

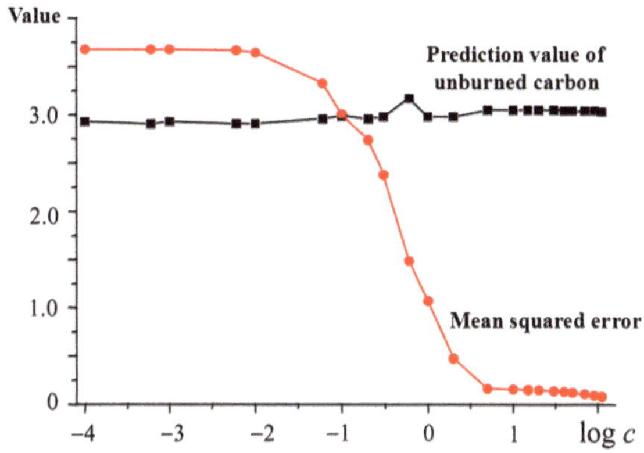

Fig. 5.7 The effects of *C* on predicting performance

Table 5.10 The parameter and capability of the model

g	*C*	Prediction value of unburned carbon	Relative error
0.0003	10	3.054	0.196%

Table 5.11 The parameter and capability of two forecast model

g	*C*	Prediction value of condition 1	Prediction value of condition 16
0.0002	10	3.65	2.06

continuous emission monitoring system. The measurements were performed a week later after the boiler switched to the pure test coal, to make the boiler conditions constant. The operating conditions during field tests are summarized in Table 5.12.

For modeling process, all the parameters of operating conditions were used as inputs and the NO_x emissions were output. Figure 5.9 presents the modeling error between the predicted and the measured NO_x emissions. The open circle presents the modeling error on the training subset, while solid inverse triangle represents the modeling error on the testing subset. The 224th case in the testing subset had the

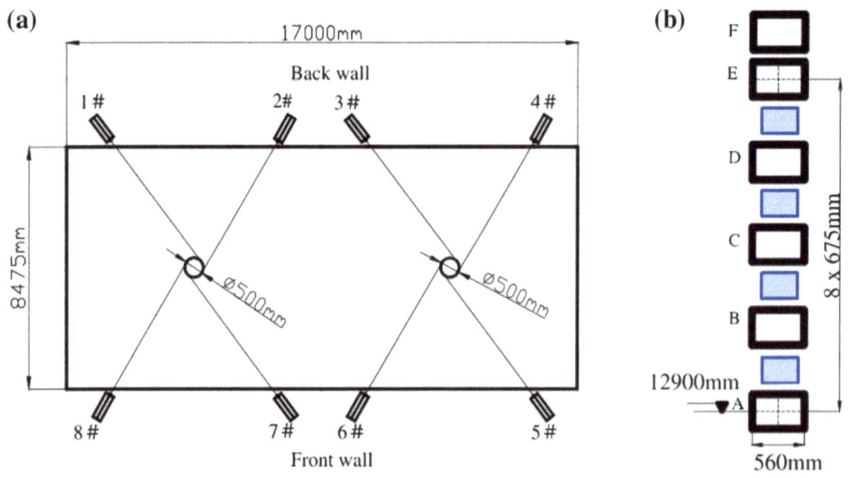

Fig. 5.8 Boiler schematic: **a** furnace cross view and **b** burner arrangement

Table 5.12 Operating conditions of the boiler during field tests

Operating conditions	Variation range
Primary air velocities (m/s)	
A level	[22.9, 29.3]
B level	[24.3, 29.2]
C level	[2.4, 28.3]
D level	[23.2, 29.1]
Secondary air velocities (m/s)	
A level	[24.7, 35.8]
B level	[23.8, 31.7]
C level	[23.4, 36.9]
D level	[20.9, 35.3]
E level	[19.7, 39.0]
F level	[2.1, 9.9]
Speed of pulverizers (rpm)	
A level	[497.7, 651.5]
B level	[237.2, 505.0]
C level	[0.0, 626.3]
D level	[300.9, 635.8]
Load (MW)	[239.6, 331.0]
Coal quality	
Volatile content (%)	31.96, constant
Moisture content (%)	11.06, constant
Ahs content (%)	13.6, constant
Low calorific value (MJ/kg)	23.49, constant
NO_x emissions (ppm)	[258.7, 407.2]

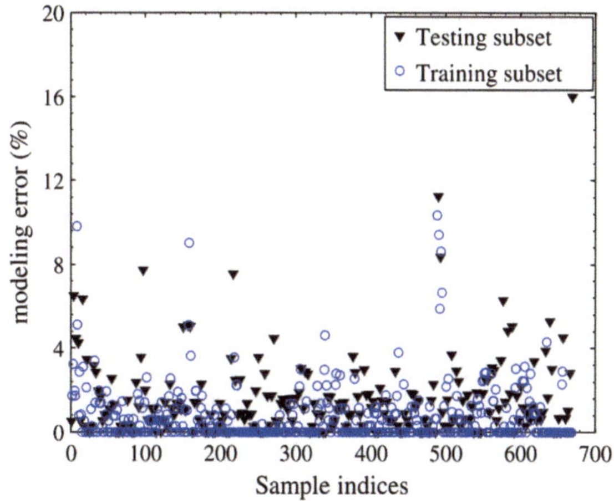

Fig. 5.9 Modeling error of the SVR model

maximum modeling error of 16%. Also, 97% (647 cases) of the total 670 cases had modeling error less than 5%. The mean modeling error and the correlation factor were 1.08 and 0.95%, respectively. As a whole, the predicted values showed rather good agreement with the measured values. The reliability and the accuracy of this SVR model have also demonstrated by in situ field tests.

5.4 Summary

In this chapter, SVM is introduced. It is a very popular method in the field of machine learning and data mining. It is based on the structural risk minimization principle and has better generalization ability than that of neural network and other machine learning methods. It is considered to be very suitable for processing small and multidimensional samples.

SVM mainly includes SVC and SVR, the principle of them is illustrated in this chapter. There are many applications of SVC and SVR in the modeling of combustion characteristics, such as coal identification (SVC), the prediction of ash fusion temperature, unburned carbon in fly ash and NO_x emission (SVR). These applications are all elaborated in this chapter.

References

1. Cortes C, Vapnik V. Support-vector networks. Mach Learn. 1995;20(3):273–97.
2. Drucker H, Burges C, Kaufman L, Smola A, Vapnik V. Support vector regression machines. Advances in neural information processing systems, 1997:155–161.

3. Chang C, Lin C. LIBSVM: a library for support vector machines. ACM Trans Intell Sys Technol. 2011;2(3):27.
4. Wang C-L, Zhou H, Li G-N, Qiu K-Z, Cen K-F. Combining support vector machine and genetic algorithm to predict ash fusion temperature. Proc CSEE. 2007;27(8):11–5 (In Chinese).
5. Wang C-L, Zhou H, Zhou Z-H, Ling Z-Q, Li G-N, Cen K-F. Support vector machine modeling on the unburned carbon in fly ash. Proc CSEE. 2005;25(20):72-76 (In Chinese).
6. Wu F, Zhou H, Zheng L-G, Cen K-F. Application of scaleable chaotic ant colony algorithm in control of unburned carbon in fly ash. J Zhejiang Univ (Eng Sci). 2010; 44(6): 1127-1131. (In Chinese)
7. Li J-Z, Zhou H, Wang C-L, Cen K-F. Employing support vector machine to predict the ash fusion temperature of coal blends. J China Coal Soc. 2007;32(1):81-84 (In Chinese).
8. Zheng L, Zhou H, Wang C, Cen K-F. Combining support vector regression and ant colony optimization to reduce NOx emissions in coal-fired utility boilers. Energy Fuels. 2008;22 (2):1034-1040.
9. Zheng L, Zhou H, Cen K-F, Wang C. A comparative study of optimization algorithms for low NOx combustion modification at a coal-fired utility boiler. Expert Syst Appl. 2009;36 (2):2780-2793.
10. Zhou H, Zheng L, Cen K-F. Computational intelligence approach for NO(x) emissions minimization in a coal-fired utility boiler. Energy Convers Manag. 2010;51(3): 580-586.

Chapter 6
Combining Neural Network or Support Vector Machine with Optimization Algorithms to Optimize the Combustion

Abstract In this chapter, optimization algorithms of GA, ACO, and PSO are introduced. They are combined with SVM and ANN and show convenience in experiments. Also multi-objective optimization is introduced, like MOCell, AbYSS, OMOPSO, and SPEA2. They are compared in many aspects. Among them, OMOPSO and MOCell are the proposed algorithms for online multi-object optimization of coal-fired boilers.

Keywords Optimization algorithm · Genetic algorithms (GA) · Ant colony algorithms (ACO) · Particle swarm algorithms (PSO) · Combining optimization NO_x emissions modeling · Low NO_x combustion · Multi-objective optimization MOCell · AbYSS algorithm · OMOPSO algorithm · SPEA2 algorithm Experimental setup

6.1 Introduction of Optimization Algorithms

6.1.1 Genetic Algorithms

6.1.1.1 Introduction to GA

GA is a popular random search technique which attempts to incorporate the idea of natural evolution. The basic concept of GA is designed to simulate processes in the natural system necessary for evolution, specifically those that follow the principles first presented by Charles Darwin in terms of the survival of the fittest. Three genetic operations: selection, crossover, and mutation are employed in GA to solve a wide range of complex optimization processes.

H. Zhou and K. Cen, *Combustion Optimization Based on Computational Intelligence*, Advanced Topics in Science and Technology in China,
https://doi.org/10.1007/978-981-10-7875-0_6

6.1.1.2 The Description of GA

An optimization problem can be described as follows:

$$\begin{cases} \max & f(X) \quad (6.1) \\ \text{s.t.} & X \in R \quad (6.2) \\ & R \subseteq U \quad (6.3), \end{cases}$$

where X is decision variable, $f(X)$ is objective function, (6.1) and (6.2) are limiting conditions, U is the basis space, R is a subset of U. X-satisfied limiting conditions is called feasible solution, R is a set composed by X, called feasible solution set.

GA codes every elements X_i in decision variable X as a genetic gene, all the possible value is called allele. In this way, X is regarded as a chromosome formed by n genes. Chromosome X is also called individual X, fitness of every individual X is calculated according to certain rule. Fitness of individual X is related to function $f(X)$. In GA, solution space is composed by decision variable X. Searching for the optimal solution is proceeded through searching chromosome X, so searching space is constituted by chromosome X.

Similar to the crossover and mutation between chromosomes, on which the nature evolution depends on, GA copies this processes in searching the optimal solution. When GA is implemented, genetic operators act on the population $P(t)$, three genetic operations are employed, then a new population $P(t + 1)$ will be generated. Three genetic operations are described as follows:

Selection. After every individual fitness is calculated, the more excellent individuals in $P(t)$ will be selected to $P(t + 1)$;
Crossover. Collocate individuals in $P(t)$ randomly, and for every individual, their chromosomes will be exchanged according to a certain crossover rate;
Mutation. Alter the certain gene loci allele value of every individual in $P(t)$ according to the mutation rate.

6.1.1.3 The Process of GA Approach

The flowchart of GA is shown in Fig. 6.1.

- Step 1 Initialization. Set evolution generation $t = 0$; Set the maximal evolution generation T; Create the initial population randomly as $P(0)$;
- Step 2 Individual assessment. Calculate every individual fitness $P(0)$;
- Step 3 Selection. Act selection operator on population.
- Step 4 Crossover. Act crossover operator on population.
- Step 5 Mutation. Act mutation operator on population to get $P(t + 1)$;
- Step 6 Terminated if satisfied. Output the optimal solution which has the best fitness.

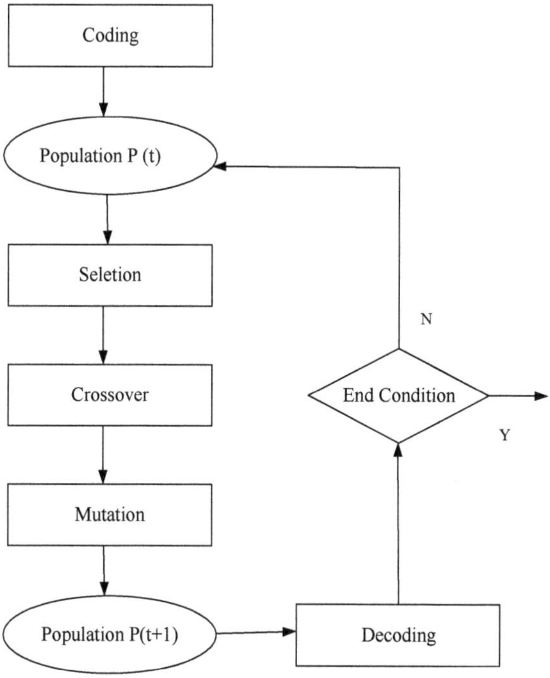

Fig. 6.1 Flowchart of genetic algorithm

6.1.2 *Ant Colony Algorithms*

6.1.2.1 Introduction to ACO

The ant colony optimization algorithm is a new metaheuristic for solving NP-hard problems, which draws its inspiration from the behavior of real ants as they moved from their nets toward a food applied. ACO has been successfully applied to solve some complex combinatorial optimization problems with NP-hard characteristic, such as traveling salesman problem, quadratic assignment problems, vehicle routing problems, and scheduling problem. Early, ant colony optimization is only used to optimize the discrete domain problem such as TSP, whereas application of ACO to continuous domain optimization is not straightforward. Currently, there are various variants of ant colony optimization for continuous domain problems.

6.1.2.2 The Description of ACO

For simplicity, the optimized problem consists only of one variable x. It is well known that any variable belonged to continuous domain can be scaled to the range of [0, 1] by Eq. (6.4) provided that the interval of the variable is [Lo, Up], where Lo and Up are the lower and upper bounds of the variable x,

$$x' = \frac{x - \text{Lo}}{\text{Up} - \text{Lo}}, \tag{6.4}$$

where x' is a scaled variable. It is easy to derive x from x' by unscaling. Any random quantity in the range of [0, 1] can be represented as a decimal fraction with d digits precision by using a bunch of character string consisting of the number.

For example, the decimal fraction 0.57895645 with 8-digit precision can be represented by a bunch of character string, 57895645. The character string is assumed to be represented as

$$d_0 d_1 d_2 \ldots d_{(l-2)} d_{(l-1)}$$

in which l is the precision. Then, the character string can be decoded into a decimal fraction through the following formula:

$$x = \sum_{i=0}^{l} d_{(i)} * 10^{-(i+1)} \tag{6.5}$$

where x is the decoded variable, $d_{(i)}$ is the ith character in the character string. Then x can be unscaled to be in the range of [Lo, Up].

The above basic idea is illustrated in Fig. 6.2, in which every node can only be the value of 0, 1, 2, …, 9. As shown in Fig. 6.2, the line connecting 8 nodes represents the 8 digits of character string, 31843356, namely the real value of the scaled variable x is 0.31843356 decoded by Eq. (12). In order to introduce the idea into ant colony optimization, some definitions are followed. The node in Fig. 6.2 is called as the city, the line connecting the nodes is referred to as the route, and each character (for d_1 in Fig. 6.2 instance) in the character string is called as a layer consisting of 10 cities. Now, the regular ACO method can be applied to optimization in the continuous domain. A variable x with d digits precision can be illustrated by a route connecting d cities. The ant must travel in a sequence in a route from one city on the first layer to a city on the last layer in Fig. 6.2. In each layer, the ant can visit anyone in these 10 cities. A completion of a run of traveling will produce a route representing a solution of the scaled variable x. As a result, the real variable can be easily derived from the scaled variable x.

Let $\tau_i(j)$ be the intensity of trail information on the edge between the current city on the ith layer and the jth city on $(i + 1)$th layer at iteration t, and use $\tau_i(j)$ to simulate the pheromone of real ants. Suppose m is the total number of ants, at iteration t each ant selects from its current layer i to city j on $(i + 1)$th layer according to the following probability distribution:

$$S_i = \begin{cases} \arg\max\limits_{j=0,1\ldots,9} (\tau_i(j)) & q \leq q_0 \\ p_i(j) = \dfrac{\tau_i(j)}{\sum\limits_{k=0}^{9} \tau_i(j)} & q > q_0 \end{cases}, \tag{6.6}$$

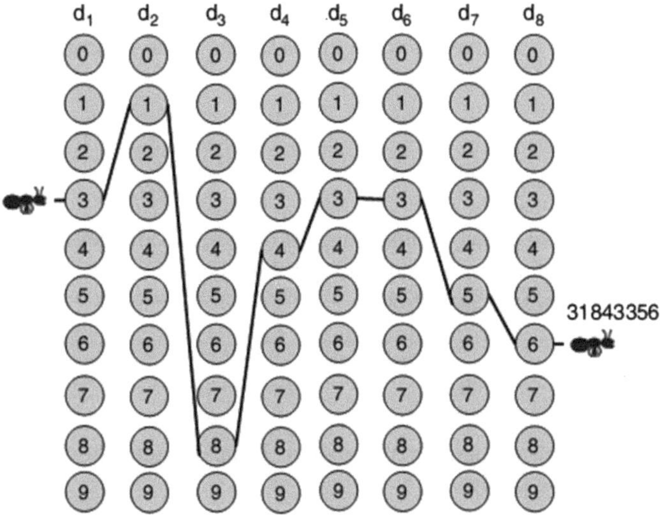

Fig. 6.2 Basic idea of ACO

where the constant q_0 is referred as to the exploitation probability factor, q is a random number. The pheromone on the edge between the current city on ith layer and the jth city on $(i + 1)$th layer is updated immediately after the edge is traveled by the ant,

$$\tau_i(j) \leftarrow (1 - \rho)\tau_i(j) + \rho\tau_0, \tag{6.7}$$

where ρ is the pheromone decay parameter, $0 < \rho < 1$, and $(1 - \rho)$ represents the evaporation of $\tau_i(j)$ between the iteration t and $t + 1$, τ_0 is the initial pheromone.

It will take an ant d steps to complete a tour traversing all the layers in Fig. 6.2. For every ant, its path traversing all layers form a solution. The value of the objective function is calculated according to the solution. The iteration best ant can be picked out by comparing m ants of solutions. If the iteration of the best ant is better than the global best ant, the global best ant will be replaced by the iteration best ant. The intensity of trail information on the route traveled by the global best ant should be changed by the updating formula:

$$\tau_i(j) \leftarrow (1 - \alpha)\tau_i(j) + \alpha/f_{\text{best}}, \tag{6.8}$$

where α is called local pheromone update strength parameter, $0 < \alpha < 1$, and $(1 - \alpha)$ represents the evaporation of $\tau_i(j)$ on the global best route, f_{best} is the objective function value given by the global best ant.

The above-presented algorithm can be very easily extended to multivariables continuous domain optimization when all variables in the optimized problem are connected in series.

6.1.2.3 Another Algorithm of ACO

It is also assumed that the adjustable range for optimized variables is represented by [Lo, Up]. A group of m ants is randomly distributed in the search space. The length of sub-range is expressed as follows:

$$\text{Len}_j = \frac{\text{Up}_j - \text{Lo}_j}{m} \quad j = 1, 2, \ldots, n, \tag{6.9}$$

where n is the dimension number of variables. Then, the initial position of ith ant is randomly derived as follows:

$$x_j = (\text{rand}(\text{Lo}_1, \text{Up}_1), \text{rand}(\text{Lo}_2, \text{Up}_2), \ldots, \text{rand}(\text{Lo}_n, \text{Up}_n)) \tag{6.10}$$

where rand(Lo_1, Up_1) is a uniformly distributed quantity in the range of [Lo_1, Up_1]. So, the initial pheromone at ith ant is represented as follows:

$$\Delta t(i) = k\mathrm{e}^{-f(x_i)}, \tag{6.11}$$

where k is a constant, x_i is optimization variables with n dimensions. Obviously, the smaller the value of the objective function $f(x)$, the more is the initial pheromone at the ith ant. This algorithm consists of the global search process and the local search process. The iteration best ant is picked out marked as a best index when all ants complete a time of iteration. In the global search process, other ants move toward the iteration best ant to find a good solution. The ith ant moves toward the iteration best ant according to a probability as defined by the following formulation:

$$p(i, \text{bestIndex}) = \frac{e^{t(\text{bestIndex}) - t(i)}}{e^{t(\text{bestIndex})}}, \tag{6.12}$$

where $t(i)$ is the pheromone at ith ant. The more the ith ant deviates from the iteration best ant, the higher is the possibility of the ith ant moving toward the iteration best ant. The new position (or solution) of the ith ant is updated by the following rules:

$$x_i = \begin{cases} x_i + \lambda(x_{\text{bestIndex}} - x_i) \\ p(i, \text{bestIndex}) < p_0 \\ x_i + \text{rand}(-1, 1) * \text{Len}_i, \text{otherwise} \end{cases} \tag{6.13}$$

where λ is called as the traverse step of other ants and p_0 is a constant, $0 < p_0 < 1$, $0 < \lambda < 1$.

Unlike the global search process applied to all ants with an exception of the iteration best ant, the local search process is only performed on the iteration best ant. It is expected that the iteration best ant will find a better solution by a random search process within a small region around the current solution. After the random search process is completed, the new position marked as tmpbest is obtained as follows:

$$x_{\text{tmpbest}} = \begin{cases} x_{\text{bestIndex}} + w \\ \text{random}(-1,1) < 0.5 \\ x_{\text{bestIndex}} - w, \text{otherwise} \end{cases} \tag{6.14}$$

$$w = w_{\max} - (w_{\max} - w_{\min}) * \frac{\text{iter}}{G} \tag{6.15}$$

where w is the traverse step of the iteration best ant, $w_{\max}$ and $w_{\min}$ are constants, generally, $1 < w_{\max} < 1.4$, $0.2 < w_{\min} < 0.8$; iter is the counter of the current cycle; G is the maximum cycle.

If the new position of the iteration best ant is better than the old position according to the pheromone defined in Eq. (6.11), the ultimate position of the iteration best ant is updated. The update rule is represented as follows:

$$x_{\text{bestIndex}} = \begin{cases} x_{\text{tmpbest}} \\ \Delta t(\text{tmpbest}) < \Delta t(\text{bestIndex}) \\ x_{\text{bestIndex}}, \text{otherwise.} \end{cases} \tag{6.16}$$

Pheromone update is a process of modifying the probability distribution used by the ants during the construction process, so that it can guide the ants toward better solutions. After all ants completed the search process just described, the update rule of the pheromone at the ith ant is represented by

$$t(i) = \rho * t(i) + \Delta t(i), \tag{6.17}$$

where ρ is the evaporation factor of pheromone. In summary, the search approach contains the following major steps:

Step 1. Set the population of ants, m and the number of the maximum cycle, G.

Step 2. Determine the adjustable range for optimized variables x, initialize the position of a group of m ants by Eq. (6.10) and the initial pheromone at ith ant by Eq. (6.11).

Step 3. Find the iteration best ant, marked as bestIndex.

Step 4. Calculate transfer probability of the other ants moving toward the iteration best ant according to Eq. (6.15). Update the position of the ith ant through Eq. (6.13).

Step 5. Update the position of the iteration best ant through Eqs. (6.14)–(6.16).
Step 6. After all ants complete a time of iteration, update the pheromone at the *i*th ant by Eq. (6.17).
Step 7. When the counter iter is less than the maximum cycle *G*, go back to step 3 and repeat the above 3–6 steps until the ending condition is met.

In this work, the first ant colony algorithm was marked as ACO^1 and the second ant colony algorithm was marked as ACO^2 in order to distinguish from each other.

6.1.3 Particle Swarm Algorithms

PSO is a stochastic global optimization technique inspired by social behavior of bird flocking. In PSO, each member of the population is called a "particle", each has two properties: position property and velocity property. To find a good solution, each particle "flies" around in the multidimensional search space with a velocity, which is constantly updated by the particle's own experience and experience of particle's neighbors or the experience of the whole swarm. Updating of each particle consists of velocity update and position update. The position update which provides a possibly better solution to optimize problem is performed through velocity update. PSO is initialized with a group of random particles with different positions' property and velocities' property. Each particle's position property is treated as a solution in a D-dimensional space. The *i*th particle's position is represented as

$$x_i = (x_{i1}, x_{i2}, \ldots, x_{iD}). \tag{6.18}$$

The previous best position of the *i*th particle that gives the best fitness value is represented as

$$p\text{best} = (p_{i1}, p_{i2}, \ldots, p_{iD}). \tag{6.19}$$

The best particle among all the particles in the population is represented as

$$g\text{best} = (g_1, g_2, \ldots, g_D). \tag{6.20}$$

Velocity, the rate of the position change for *i*th particle is represented as

$$V_i = (V_{i1}, V_{i2}, \ldots, V_{iD}). \tag{6.21}$$

In every iteration, the velocity and position updating rule for each particle is given as

$$V_{id}^{t+1} = \omega V_{id}^{t} + c_1 r_1 (pbest_{id}^{t} - x_{id}^{t}) + c_2 r_2 (gbest_{d}^{t} - x_{id}^{t}) \tag{6.22}$$

$$x_{id}^{t+1} = x_{id}^{t} + V_{id}^{t+1}, \quad i = 1, 2, \ldots, n \tag{6.23}$$

where V_{id}^{t+1} is the velocity property along the dth dimension of particle i in iteration $t + 1$, while x_{id}^{t+1} is the position property along the dth dimension of particle i in iteration $t + 1$, n is the number of particle or size of the population. c_1 and c_1 are constants named acceleration coefficients, r_1 and r_2 are two independent random numbers uniformly distributed in range of [0,1]. The velocity V_i for each particle is a random number within $[-V_{max}, V_{max}]$, where V_{max} is a problem-dependent constant defined in order to clamp the excessive roaming of particles.

In the present study, V_{max} is equal to the one-tenth of the length of the optimized variables, say, the upper bound subtracted by lower bound. $pbest_{id}^{t}$ is the previous best position along the dth dimension of particle i in iteration t; $gbest_{d}^{t}$ is the precious best position among all the particles along the dth dimension in iteration t, where ω is defined as inertia weight factor. An empirical study of PSO with inertia weight has shown that a relatively large ω has more global search ability while a relatively small one results in a faster convergence.

In the original PSO, the control parameters include the inertia weight ω, acceleration coefficients c_1 and c_2. It is not a trivial task for users to select appropriate values of these parameters. In most cases and/or applications, the best parameters for the PSO algorithm are determined in an empirical manner. In addition, the tuning of the algorithm parameters is specific for the case studied. Therefore, few-parameters algorithm is preferable for use. Clerc and Kennedy proposed an improved update rule for particle's velocity based on convergence analysis.

$$\begin{aligned} V_i(t+1) &= \chi(V_i(t) + r_1 c_1 (pbest_i - X_i) + r_2 c_2 (gbest_i - X_i)) \\ \chi &= \begin{cases} 1, & \text{if} \quad \phi = (c_1 + c_2) \le 4 \\ \frac{2}{\left|2-\phi-\sqrt{\phi^2-4\phi}\right|} & \text{if} \quad \phi > 4 \end{cases} \end{aligned} \tag{6.24}$$

Clerc and Kennedy recommended $\chi = 0.729$ and $c_1 = c_2 = 2.05$. Values of $\chi = 0.6$ and $c_1 = c_2 = 2.83$ were proposed by Trelea through the convergence analysis and a better performance was achieved when compared to those by Clerc and Kennedy. The main advantage of the above update rules is that the number of control parameters for algorithms is very small. To make the computation more efficient, parameters selection in a trial and error manner should better be avoided. A few-parameters algorithm is preferred. Therefore, the methods by Clerc and Kennedy or by Trelea rather than those by choosing the optimal parameters through time-consuming attempts were tried here to check its applicability to low NO_x emissions combustion optimization.

6.2 Combining Neural Network and GA to Optimize the Combustion

Chinese version: % % ANN和GA相结合的算法案例%%
English translation: Some cases about the combination of artificial intelligence neural network and genetic algorithm.

6.2.1 Experiments

The experiments have been carried out at a 300 MW, tangentially fired dry bottom boiler with a large furnace of 14.022 × 12.33 m^2 section and height of 48.5 m. The tilting fuel and combustion air nozzles including five primary air nozzles and six secondary air nozzles and one over fire air (OFA) nozzle are located in each corner of the furnace. All nozzles can be tilted in a vertical direction over ca. ±20 °C from the horizontal axis, both upwards and downwards to adjust the reheated steam temperature for the varying fouling conditions of the furnace. Four levels of the fuel burners are put into operation under the rated load. The medium-speed coal pulverizers are employed to supply the coal–air mixture to the burners on the corresponding levels. The dimensions of the furnace and the arrangement of the burners are illustrated in Fig. 6.3. A total of 21 tests were performed on this boiler.

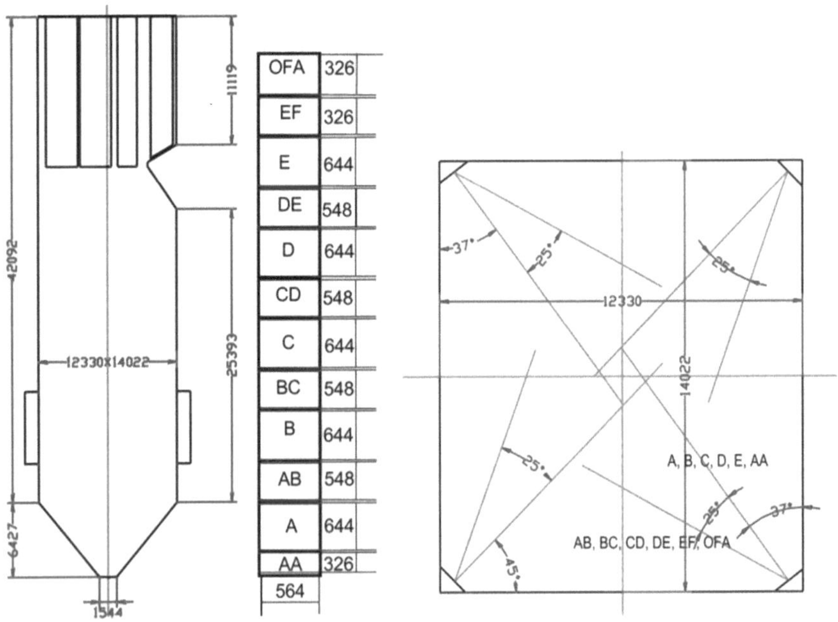

Fig. 6.3 The sketch of the investigated utility boiler

The operating parameters of the tests are listed in Tables 6.1 and 6.2. Six coals are burned in the tests, and proximate analyses of the coals are also listed in Table 6.2. Note that the damper position indicated in Tables 6.1 and 6.2 is the percentage opening of a butterfly valve across the area of the given inlet port. Field tests show that the air is almost evenly distributed between the same level of the burners in proportion to the flow areas according to the percentage damper opening.

Tests 6-1 and 6-2 are conducted to test the effect of the secondary air distribution pattern on the carbon burnout. Tests 3 and 4 evaluate the effect of OFA. Test 5, 12, and 6 investigate the influence of fuel air and burner tilt, respectively. In tests 7–9, the oxygen concentration in the flue gas is taken as the primary variable of the tests. The effect of pulverizer operating mode on the carbon burnout is studied in tests 10–11. Tests 14–17 and tests 18–21 are the experiments carried out under the low furnace load and the medium load, respectively, investigating the effects of changing load, the oxygen concentration in flue gas, and the pulverizer operating mode. Test 13 is the optimal operating condition summarized in the tests, in which a lower carbon burnout is achieved.

During the experiments, coal fineness was kept constant, and main gas species concentrations were continuously monitored in the boiler outlet prior to the air heater.

Table 6.1 The boiler operating conditions

Case	Load (MW)	The secondary air burner damper position (% open)					The feeding rate of the coal feeder (*t*/*h*)				
		AB	BC	CD	DE	EF	A	B	C	D	E
1	294.1	100	100	70	70	100	0	27	27	25	26
2	295.4	100	100	100	70	60	0	27	27	25	26
3	294.3	100	100	100	100	100	0	27	27	25	26
4	296.1	100	100	100	100	100	0	26.4	25.8	25.8	25.8
5	294	100	100	100	100	100	0	26	27	26	25
6	294.8	100	100	100	100	100	0	26.5	26	25.9	25.5
7	295.4	24	100	100	100	100	0	25	26	26	26
8	291.6	44	100	100	100	100	0	25	26	26	26
9	292.8	45	100	100	100	100	0	25	26	26	26
10	296	95	100	100	100	4	27	27	26	26	0
11	299.5	100	66	42	100	100	26	27	0	26	25
12	295.6	83	71	47	86	78	26	27	0	27	25
13	298.6	100	100	0	100	100	28	25	0	27	27
14	213.7	70	70	70	70	70	19	0	18	19	19
15	216.1	70	70	70	70	70	19	0	19	19	20
16	216.8	70	70	70	70	70	19	0	19	18	19
17	215.3	70	70	70	70	70	19	18	20	20	0
18	251.3	100	100	100	94	40	24	19	23	22	0
19	251	100	100	100	100	30	24	20	23	22	0
20	250.9	100	100	100	100	30	24	20	24	21	0
21	257.6	65	99	99	99	99	0	21	25	23	22

Table 6.2 The boiler operating conditions

Case	O_2 (vol %)	Damper position (% open)		Pressure drop (kPa)	The coal properties				Burner tilt (%)	Primary air pressure (kPa)
		Fuel air	OFA		V_{ar} (%)	A_{ar} (%)	M_{ar} (%)	$Q_{net,ar}$ (kJ/kg)		
1	3.54	100	85	0.71	28.92	11.552	8.1	25,350	50	7.8
2	3.47	100	83	0.75	28.92	11.552	8.1	25,350	50	7.52
3	3.05	80	0	0.796	28.26	17.019	7.87	23,364	100	7.58
4	3	79	84	0.656	28.26	17.019	7.87	23,364	100	7.84
5	2.88	20	84	0.795	28.26	17.019	7.87	23,364	100	7.82
6	3.14	47	83	0.74	29.32	13.076	11.57	22,222	20	7.69
7	2.09	81	87	0.555	29.94	11.522	11.6	23,820	100	7.45
8	3.67	81	82	0.78	29.94	11.522	11.6	23,820	100	7.25
9	4.02	83	82	0.82	29.94	11.522	11.6	23,820	100	7.5
10	3.4	50	88	0.572	27.17	12.798	11.1	22,636	20	7.52
11	3.14	50	85	0.67	27.17	12.798	11.1	22,636	20	7.17
12	3.09	20	87	0.765	27.17	12.798	11.1	22,636	20	7.6
13	2.35	80	90	0.565	27.17	12.798	11.1	22,636	50	7.6
14	4.27	38	34	0.56	26.225	17.436	11.7	21,810	100	6.9
15	3.68	38	33	0.43	26.225	17.436	11.7	21,810	100	6.8
16	5.05	38	31	0.79	26.225	17.436	11.7	21,810	100	7.58
17	4.45	38	33	0.5	26.225	17.436	11.7	21,810	100	7.1
18	3.89	85	60	0.5	26.225	17.436	11.7	21,810	50	8.3
19	3.08	85	58	0.4	26.225	17.436	11.7	21,810	50	8.2
20	4.8	85	57	0.67	26.225	17.436	11.7	21,810	50	8.2
21	4.07	85	58	0.7	26.225	17.436	11.7	21,810	100	7.5

Solid samples were withdrawn from the flue gas by an isokinetic cyclone dust collector. The carbon burnout data reported here are average values over several hours of stable operation. Pulverized coal samples were extracted from the burner supply lines with isokinetic multi-point sampling system. The measurements were performed a week later after the boiler is switched to the pure test coal, to make the boiler conditions constant. Only pure test coal was fired during a test. The measured unburned carbon in fly ash is summarized in Table 6.3.

6.2.2 *Result and Discussions*

Artificial neural network training results in the achievement of the values of connecting weights between processing elements in the input and hidden layer and

Table 6.3 The measured and estimated unburned carbon in fly ash

Case	1	2	3	4	5	6	7	8	9	10
Measured	4.0	3.3	8.2	7.4	4.6	5.5	5.5	4.2	2.9	4.6
Estimated	4.0	3.3	8.2	7.4	4.6	5.5	5.5	4.1	3.0	4.6

Case	11	12	13	14	15	16	17	18	19	20	21
Measured	3.1	2.8	2.4	1.9	2.3	1.6	1.6	2.3	2.9	1.5	3.9
Estimated	3.1	2.8	2.4	2.0	2.4	1.6	1.6	2.3	2.9	1.5	3.9

between the hidden layer and output layers which minimize the differences between the network output and the measured values. Experimental data stated above are used to find the relation between the coal qualities, operational parameters, and the carbon burnout. The following variables are chosen as inputs to the network.

(1) Variables characterizing the operational conditions of the boiler (boiler load, secondary air burner damper position, the feeding rate of coal feeder, OFA port damper position, fuel air port damper position, pressure drop between the wind box and the furnace, oxygen concentration in the flue gas, primary air flow rate, nozzles tilt).
(2) Variables characterizing the coal properties [content of ash (% as received), volatile (% as received), moisture (% as received), and heat value (kJ/kg as received)]. The output was the unburned carbon in fly ash. Using an artificial neural network with 21 input neurons in the input layer, 1 output neuron in the output layer, and 24 hidden neurons in the hidden layer to model the carbon burnout characteristics of the boiler. The 21 input variables are illustrated in Tables 6.1 and 6.2.

Training the neural network is an important step for developing a useful network. The experimental data of tests 1–12 and tests 14–21 are used as the learning samples to train the neural network. The learning parameter is 0.9, and the moment attachment and the self-adaptive learning step size methods are adopted in the training process. The training will stop when the mean square error of the system is less than 0.00001. As a result, a total of 58,508 steps are needed to achieve the correct weight and threshold values.

Using the data of test 13 as input to verify the accuracy of the model, the output of ANN is 2.423%, while the experimental result is 2.35%. Thus, the absolute error is 0.073% and the relative error is 3.125%. Such an examination was also done on the training samples, and the predicted outputs of the artificial neural network are also listed in Table 6.3. The comparison of measured data and the estimated data of the model are illustrated in Fig. 6.4, in which one can observe that the accuracy of model for the learned samples is very good and the error increases slightly for the unlearned sample of test 13.

The neural network simulator developed in this work illustrates the ability of modeling the complex coal combustion performance in a tangentially fired boiler. Compared with the computational fluid dynamics methods, the neural network may

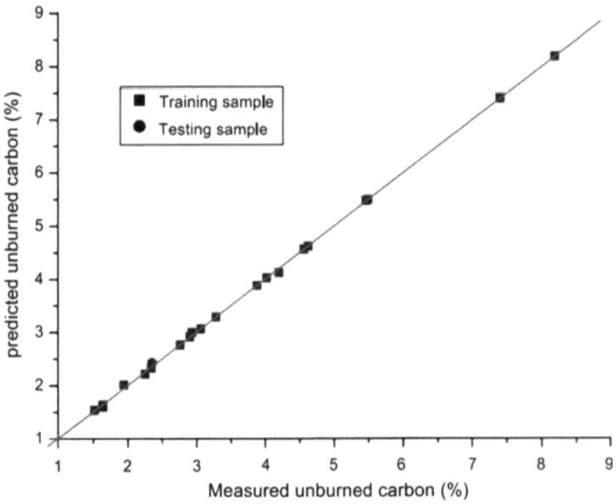

Fig. 6.4 The effect of the neural network model

not directly contribute to a better understanding of the complicate combustion phenomena in boilers, but the artificial neural network gives the engineers an alternative approach to evaluate the combustion performance of the boiler. It is more convenient and direct, consuming much less time than the CFD methods. The neural network takes a lot of variables into account, and it may identify which variables may be significant for the combustion process, which is difficult for the estimation of combustion performance by empirical relations.

The accuracy and robustness of an artificial neural network are significantly affected by the availability of the training data. When the amount of the training data is limited or the accuracy of the data is not so good, the neural network performs significant errors. In this case, the accuracy of experimental data is reliable and can be verified by repeating tests. But it is noted that neural networks provide an approximation to a nonlinearity and can be only trusted within the scope of the training data. To overcome the problem of limited training data available, stacked neural network to improve the accuracy and robustness of an artificial network model is presented by Wolpert. Results have shown that improved predictions can be obtained using stacked neural networks.

On the other hand, because of the expensive price of measurement of the combustion characteristics and the limitations of time and number of instruments and skills required in field test, it may not be effective to carry out a lot of field tests to develop an artificial neural network model. But with the wide use of distributed control system (DCS) in the power station, it is very convenient to achieve the operational parameters from the DCS. If the online flue gas analysis recorder and the unburned carbon in fly ash monitor are installed in the boiler, a lot of data can be achieved to train the neural network. Then, the neural network can be easily reestablished to model the current combustion process in the furnace with the newest operational data. If a sufficient amount of good quality operational data are available to train and test, the neural network models for predictions of carbon burnout performance can be developed with good accuracy and robustness.

Another advantage of the artificial neural network model is that the combination of neural network and the genetic algorithms can achieve the optimum operating conditions, resulting in the highest heat efficiency of the boiler. As generally known, heat efficiency can be described as 100 minus the sum of the heat losses expressed in percent, which means the reduction of heat losses leads to a higher heat rate. The heat losses in boiler consist of heat loss due to unburned combustible, flue gas losses (including the sensible heat in dry flue gas, the sensible heat in the moisture, the sensible heat in the H_2O in the fuel and the latent heat of the moisture in the fuel), radiation loss from the unit setting.

According to the ASME standards, boiler efficiency can be obtained if the coal analysis data, the ambient environment data, the unburned carbon, the excess air coefficient, and the value of flue gas temperature leaving unit are available. If the relation between the heat losses and the coal quality, operating parameters can be expressed, then the fitness function of the genetic algorithms can be used to obtain the optimal operating conditions under various ambient conditions or coal quality. But because of the complex mechanism in the combustion process, the relation between the unburned fuel loss and the operating parameters cannot be expressed analytically, which is not similar to other heat losses.

Using the neural network to model the carbon burnout characteristic of the coal-fired boiler, the function between input operating parameters and the carbon burnout, then can be determined after the training phase of the neural network. If one considers the carbon burnout function expressed by the neural network as the fitness function, then the optimum operating parameters leading to the best carbon burnout can be achieved under various conditions by using the GA.

Since the heat efficiency of the boiler is influenced not only by the unburned fuel loss, but also the other losses, the task of the operator is to obtain the highest heat efficiency. Therefore, it is more suitable for employing boiler efficiency as the fitness function of the GA. Consequently, the search process will result in the optimum operating conditions leading to the highest boiler efficiency. In the optimizing process, the unburned carbon loss is expressed by the artificial network model. Other heat losses are calculated by ASME standards, so the relationship between the boiler efficiency and the coal qualities and operating parameters can be established in the fitness function.

Under a determined unit load, the most relevant parameters for the operators are the secondary air and the OFA damper opening position, the oxygen concentration in the flue gas. The other input parameter *s*, such as pulverizer feeder opening value, air flow rate through the pulverizer and the pressure drop between the wind box, are usually not considered as adjustable factors for combustion optimizing purpose. To illustrate an example, a practical operating condition is listed in Table 6.4. The five secondary air damper opening positions, one overfire air damper opening position, and the oxygen concentration in the flue gas is considered as the parameters to be optimized. The other parameters can also be optimized with the same method if necessary.

Table 6.4 A practical operating condition

Load	Ambient	The secondary air burner damper position (%)					The feeding rate of the coal feeder (*t*/*h*)				
		AB	BC	CD	DE	EF	A	B	C	D	E
298.7	21.35	100	100	100	100	100	26.6	26.5	0	26.5	27

O_2 (vol%)	Damper position (%)		Windbox pressure drop (kPa)	The coal properties				Burner tilt (%)	Primary air pressure (kPa)	Exit flue gas temp (°C)
	Fuel air	OFA		V_{ar} (%)	A_{ar} (%)	M_{ar} (%)	$Q_{ar,net}$ (%)			
3.515	100	88.9	0.829	27	17.5	10.0	23,750	41.875	8.57	121.4

Table 6.5 The optimized results using the genetic algorithms

Parameters	The air nozzles damper opening position (% open)						O_2 in the flue gas (vol%)	Boiler heat rate	Unburned carbon in fly ash (%)
	AB	BC	CD	DE	EF	OFA			
Scope	70–100	70–100	70–100	70–100	70–100	70–100	2.0–3.5	93.53	1.52
Value	70.84	70.27	98.41	99.31	70.54	95.65	2.0		

The adjustable range for the secondary air and overfire air damper opening position is 70–100%; based on the operating routine and security considerations, it means that the search space for the GA is determined. The adjustable range for the operating parameter can be easily regulated if needed.

A practical operating condition when burning various coals is illustrated in Table 6.5. Genetic algorithms use three various sets of calculating parameters to achieve the optimum controllable parameters for the highest boiler efficiency, under such a condition. The first calculating process chooses a population size $\mu = 50$, a crossover probability $p_c = 0.8$ and a mutation probability $p_m = 0.15$. The second process uses a population size $\mu = 50$, a crossover probability $p_c = 0.8$, and a mutation probability $p_m = 0.3$. The third process uses a population size $\mu = 50$, a crossover probability $p_c = 0.\ 2$, and a mutation probability $p_m = 0.1$. A total of 500 generations are calculated in each optimizing process.

The boiler efficiency calculated by GA under various generations is shown in Fig. 6.5. It illustrates that the search process is progressive and the rate of convergence is very fast, consuming a few minutes of CPU time on a modern desktop computer. After convergence, the results using different calculated parameters are almost similar. The optimal operating parameters, boiler efficiency, and the unburned carbon in fly ash are listed in Table 6.5. The boiler efficiency and the unburned carbon are in good agreement with the experimental data under the optimized operating conditions.

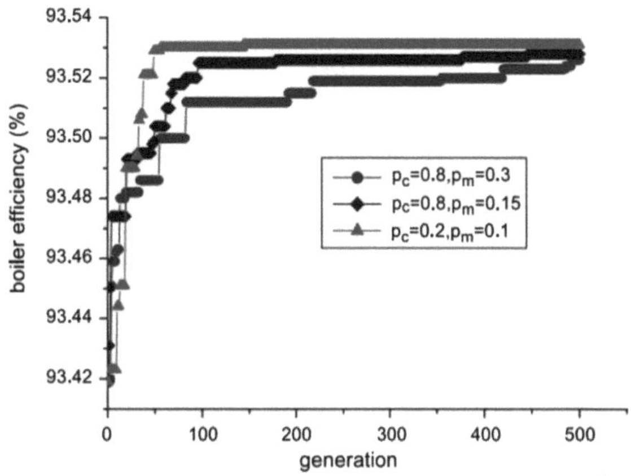

Fig. 6.5 The optimizing process of the GA

6.2.3 *Conclusions*

In this work, the influences of coal quality and operating parameters on the carbon burnout are experimentally investigated. A neural network with one input layer, one hidden layer, and one output layer was capable of "learning" the relationship between the coal quality, operational parameters, and the combustion performance. The estimated outputs of the neural network model agree well with the experimental values. It offers an alternative for engineers to evaluate the effect of operational parameters on carbon burnout behavior when coal switching happens. Although the neural network does not give any information about the mechanism that governs the simulated process, it may identify which input variables may be significant for the process. With the wide use of DCS and online flue gas analysis recorder and the unburned carbon monitor, the training process of the neural network modeling the combustion is very convenient, which gives the operators a useful tool to evaluate the combustion behavior under various operating conditions. Operation optimization based on the neural network and genetic algorithms is developed and verified in this work, resulting in the highest boiler efficiency.

6.3 Combining SVM and Optimization Algorithms to Optimize the Combustion

Chinese version: %%仅带有模型参数优化,没有运行参数寻优的SVR案例%%
English translation: SVR case in which the model parameters are optimized and the processing parameters are not optimized.

6.3.1 Modeling NO_x Emissions by SVM and ACO with Operating Parameters Optimizing

As we know, the main purpose of finding optimum parameters pair (C, γ) is to obtain good predictive accuracy according to predetermined evaluation criteria. Therefore, the main idea of ACO is to optimize the evaluation criteria as a fitness function by selecting proper parameters pair (C, γ). Figure 6.6 illustrates the algorithm process of the selection of SVR model parameters based on ACO.

6.3.1.1 Experimental Setup and Data Analysis

For the purpose of this study, NO_x emissions data from an actual power plant were used. The experiments were carried out on a 300 MW tangentially fired dry bottom boiler manufactured by Shanghai Boiler Co. Ltd. in the 1980s with large dual

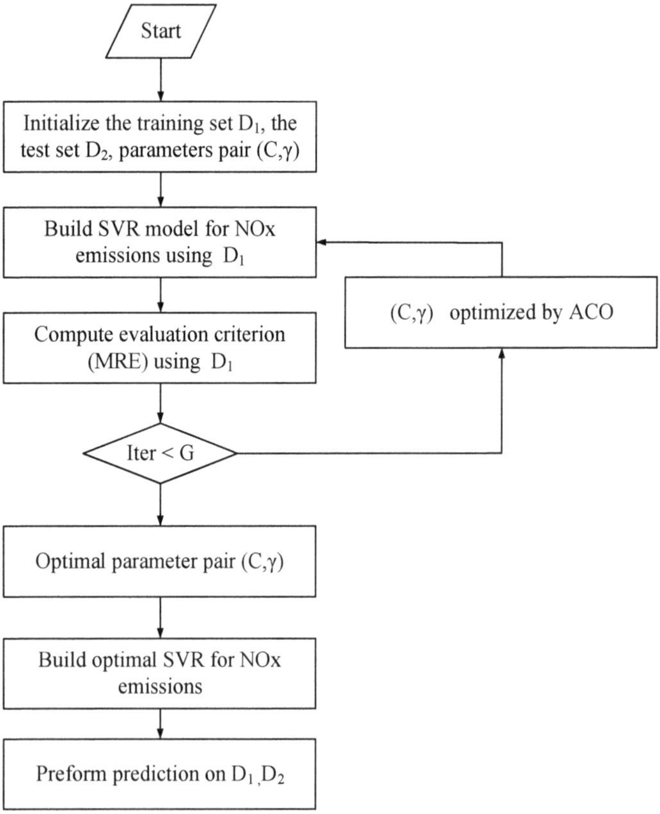

Fig. 6.6 Selection of (C, γ) by ACO

furnaces of 17 × 8.475 m^2 section and 45.5 m high, which is one unit of JianBi power plants located at the Jiangsu Province of China. The tilting fuel and combustion air nozzles including 4 primary air burners and 6 secondary air burners were located in each corner of the furnace. The fuel and primary air streams were directed at the circumference of an imaginary circle of 500 mm diameter at the center of the furnace. Coal for combustion was supplied by 4 medium-speed coal pulverizers. Coal–air mixtures were fed to the burners on A–D levels. Operation experiences showed that NO_x emissions were high due to the old-design technology of the boiler.

Bituminite was burned in this boiler. The characteristics of the combusted coal were as follows: the volatile, the ash content, moisture content, and the heating value as received were 31.96, 11.06, 13.60 wt%, and 23.49 MJ/kg, respectively. NO_x and O_2 concentrations were monitored continuously by Continuous Emissions Monitoring System (Rosemount, Emerson process management) in the boiler outlet prior to the air heater. The NO_x concentrations reported here were average values over several hours of stable operation, and they were obtained under dry gas conditions. The measurements were performed a week later after the boiler switched to the pure test coal to make the boiler conditions unchanging. Six hundred and seventy tests were carried out on this boiler, changing the boiler parameters, to analyze the characteristics of NO_x emissions of the tangentially fired system. The parameters range regulated in experiments are shown in Table 6.6. Due to the air duct arrangement restriction, only a very narrow air velocity change range of F level of secondary air can be achieved. As shown in Table 6.6, the boiler operated in a wide range, and the measured NO_x emissions and operational parameters provided a rich enough dataset for developing empirical models. The monitored NO_x emissions are shown in Fig. 6.7.

The dataset consisted of 670 cases with 22 dimensions, in which 21 inputs to the utility boiler included four levels of primary air velocity, six levels of secondary air

Table 6.6 The parameters range experienced in experiments

	Range
Boiler load (MW)	239.64–331.03
Primary air velocity (A–D level) (m/s)	2.39–29.28
Secondary air velocity (m/s)	
A–E level	19.68
F level	2–10
O_2 content in flue gas (%)	2.6–4.62
Speed of mills (rpm)	0.91–651
Total air flow (t/h)	239.64–331.03
Coal quality	Unchanged
NO_x emission	258.7–407.22

MW: Million watts; m/s: meter per second; %: one percent; rpm: revolutions per minute; t/h: tons per hour; ppm: percent per million

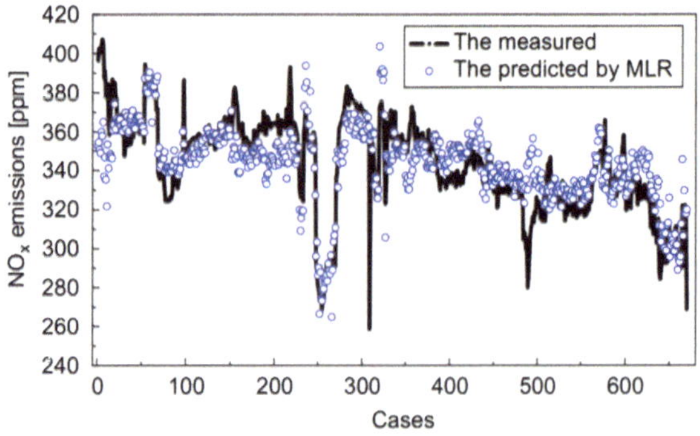

Fig. 6.7 Monitored and predicted NO_x emissions by MLR for the coal power plant

velocity, rotation speed of four mills, the boiler load, the oxygen concentration in flue gas, total combustion air flow rate and coal quality such as the volatile matter, the ash content, the heating value, and the moisture content, while the only output was NO_x emissions. According to the physical analysis, the NO_x emissions mainly depended on the above-mentioned 21 inputs of the boiler. Therefore, the SVR was used in this work to build the relation between the output of the boiler and its 21 inputs.

Sometimes, as a backup monitoring system for replacing hardware-based CEMS, a mathematical model is used to estimate the NO_x emissions. That currently used is a linear one, in the following form (Graziani et al. 2004):

$$C_{NO_x} = \sum_{i=1}^{21} w_i x_i + b_0,$$

where C_{NO_x} is the content of NO_x emissions, x_i is the ith input of the boiler and b_0 is the intercept. In this study, the total number of boiler inputs is 21, each one is characterized by a relative importance weight w_i, determined by least quadratic squared regression. Coefficients of the regression equation are given in Table 6.7. The predicted NO_x emissions using the multiple linear regression equation (MLR) are also shown in Fig. 6.7, as represented by open circles. The results show that the MLR model gave the maximum relative error of 37.7%, mean relative error (MRE) of 3.4% and the correlation coefficient of 0.77. This result derived from the linear model is not very accurate due to the inherent nonlinear relation between the boiler inputs and the output, and the further improvement in predictive accuracy can be made by resorting to a nonlinear model such as SVR and ANNs.

To eliminate the unit influence of various operational parameter of the boiler, some necessary prepossessing of the raw data before feeding them into the SVR model is needed. In this study, all the feature elements and the target values were scaled so that they fall into the range of [−1, 1]. When using these models, the computed target value should be converted back into the same scales that were used for the original target values.

Table 6.7 Coefficients of MLR for NO_x emissions

Inputs	Primary air velocity				Secondary air velocity						O_2
	A	B	C	D	A	B	C	D	E	F	
Coefficients	−2.44	−0.93	1.93	−3.07	2.98	8.88	3.63	−4.99	−1.61	2.72	6.54
	Speed of mills				Load	Total air flowrate	Coal quality				b_0
	A	B	C	D			A_{ar}	V_{ar}	$Q_{ar,net}$	M_{ar}	
	0.09	0.018	0.16	−0.07	−0.76	2.99	0.00	8.31	0.00	0.00	0.00

6.3.1.2 Results

In this section, the prediction results of NO_x emissions in the studied coal-fired utility boiler through SVR are presented. The emphasis is played on the selection of model parameters, the predictive accuracy, and the time response of the SVR model. Regarding the selection of two parameters (C, γ), two techniques were employed in this investigation. One method is the optimization approach based on Evolutional Algorithms ACO, and another method is the grid search method for comparison purpose.

6.3.1.3 Prediction Results of ACO–SVR

In the optimization approach, MRE on test set D_2 was employed as the objective function, as shown in Fig. 6.6. That means the MRE was employed to evaluate the predictive accuracy of the SVR model. In this study, all calculations were performed on a 2.4 GHz Intel Pentium IV PC with 1.5 GB RAM under Windows XP.

To begin with, a population $m = 50$ and the maximum generations $G = 100$ were chosen as the ACO's parameters. It took nearly 5053 s to get the best parameters with the MRE of 1.5905%, the optimal pairs of (C, γ) was (115.26, 0.65), as shown in the last rows in Table 6.8. Obviously, it is a time-consuming process and not suitable for the on-line built SVR model. It may typically take several seconds to train a SVR model using the training set D_1 consisting of 446 samples with 21 inputs and one output. On average, ACO with 50 initial ants and a maximum generations $G = 100$ will try at least 5000 SVR models. The searching process is extremely CPU demanding. It is straightforward to shorten the computing time required for one ACO optimization procedure by reducing the population m and the maximum generations G. However, the performance of ACO optimization procedure depends significantly on these two parameters. Therefore, the trade-off between the time cost and the performance must be examined. Results derived from various combinations of the population m and the maximum generations G investigated in this work are listed in Table 6.8. Each combination was repeated five times to minimize the results variability. Table 6.8 also gives the optimized parameters pair (C, γ), the computing time, and the model performance such as the correlation coefficient (R) and MRE on the training set D_1 and test set D_2,

Table 6.8 Results of ACO–SVR under various combinations of (m, G)

(m,G)	R		MRE (%)		(C, γ)	Computing time (s)
	D_1	D_2	D_1	D_2		
(5,5)	0.9635	0.935	0.5792	1.6172	(142.77,0.92)	25
(5,10)	0.9588	0.9355	0.6903	1.6105	(127.60,0.60)	51
(5,20)	0.9669	0.9342	0.4899	1.6084	(142.45,0.61)	100
(5,30)	0.9629	0.9355	0.602	1.5966	(116.52,0.66)	164
(5,50)	0.9632	0.9353	0.5983	1.5927	(113.84,0.63)	254
(10,5)	0.9668	0.9342	0.4784	1.6139	(124.02,0.47)	57
(10,10)	0.9649	0.9343	0.5368	1.6111	(183.56,0.62)	110
(10,20)	0.963	0.9357	0.603	1.5961	(120.28,0.62)	218
(10,30)	0.9623	0.9354	0.6172	1.5939	(115.12,0.63)	333
(10,50)	0.9626	0.9354	0.6103	1.5933	(123.76,0.60)	517
(50,100)	0.9635	0.9353	0.5889	1.5905	(115.26,0.65)	5053

respectively. The typical MRE and parameters pair (C, γ) histories optimized by ACO are shown in Figs. 6.8 and 6.9, in which the population $m = 5$ and the maximum generations $G = 50$ were employed. Obviously, the ACO can find the best parameters pair (C, γ) within a few iterations when the MRE on test set D_2 was employed as the objective function. It can be concluded that the algorithms have very fast speed.

As we know, the computing time spent to execute the optimization procedure is mainly dependent on how many times the SVR model is tried and trained. Obviously, the product of the population and the maximum generations, ($m \times G$), can be considered as the times the SVR model is recalled. The performance of SVR model characterized by MRE is plotted as a function of the product, as illustrated in Fig. 6.10. Results show approximately monotonic relations between the MRE with the product, ($m \times G$). Especially, the computing time is perfectly linearly correlated with the product, ($m \times G$). Larger products led to larger computing times. However, the MRE of SVR model will reach a plateau when the product is larger than 150. So, the trade-off must be made between the performance and the computing time. As for this study, the product of 150, for example, the combination of ($m \times G$) = (5, 30), will be suitable and sufficient.

After the ACO was applied to search for the optimal parameters sets, the SVRs prediction model was built. To check the generalization property of the model, another dataset D_2, which had not been used in training, was employed to validate the models. The graphical comparisons between the actual and the predicted NO_x emissions are shown in Figs. 6.11 and 6.12. The absolute error and the relative error are very easy to calculate for all cases from the figures. As for the training set D_1, the maximum absolute error was about 48 ppm, while the maximum relative error was 18.56%. As for the test set D_2, the maximum absolute error was about 41 ppm, while the maximum relative error was 15.99%. 92% of cases in test data D_2 had the relative error smaller than 5%. The results are better than the reported 20% error [1] and are comparable to the 5% error (Azid et al. 2000).

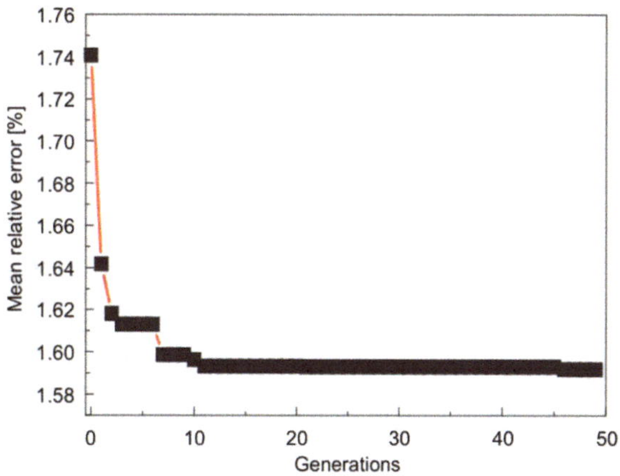

Fig. 6.8 Typical mean relative error history optimized by ACO

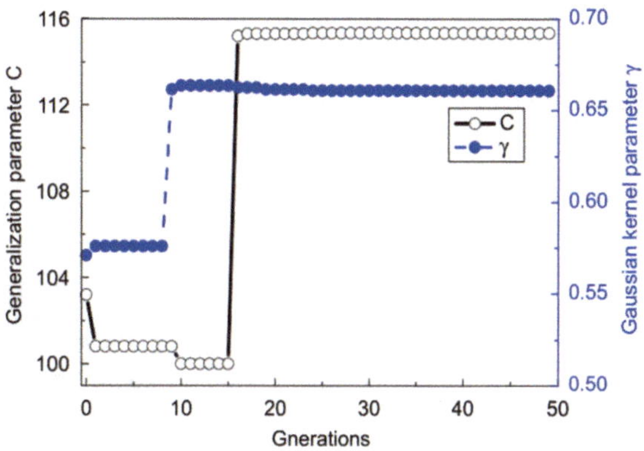

Fig. 6.9 Typical C and γ history optimized by ACO

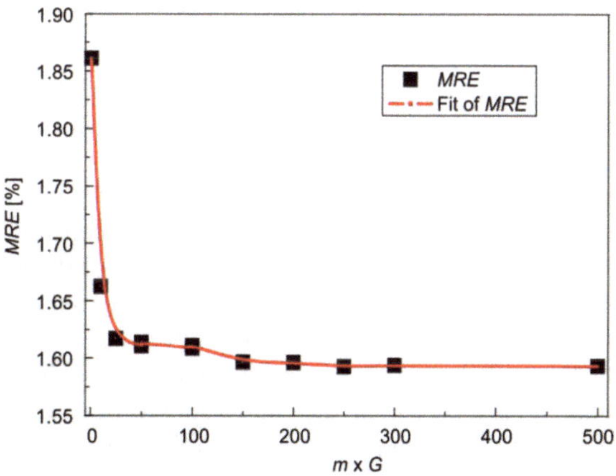

Fig. 6.10 Dependence of MRE on the $m \times G$

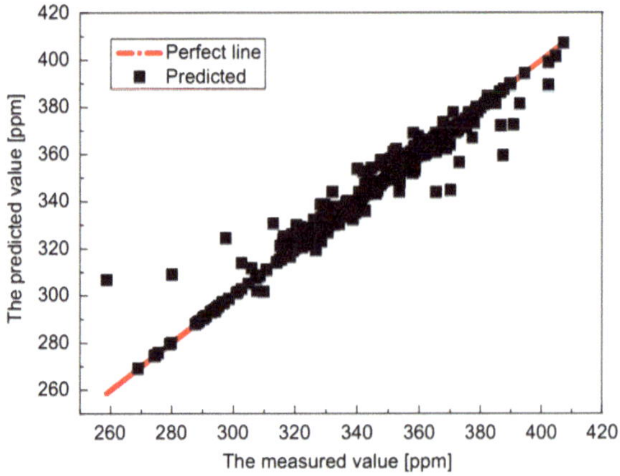

Fig. 6.11 Predicted NO_x emission on training set D_1 and from ACO–SVR

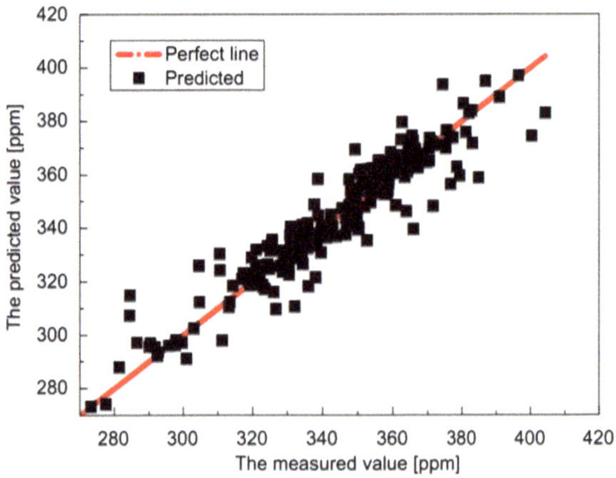

Fig. 6.12 Predicted NO_x emission on test set D_2 from ACO–SVR

This model produced a MRE error of 0.6020% for the training set D_1, 1.5966% for the test set D_2 and 0.9388% for the whole set, the corresponding correlation coefficients (R) were 0.9629, 0.9355, and 0.9537. It can be concluded that the predicted values were in very good agreement with the experimental values. 164 s of computing time is acceptable for online constructing the model for NO_x emissions, which is expected in the development of PEMS and combustion optimization software package. In addition, the relatively small standard deviation of the performance for the SVR model indicates that *i* is quite robust.

For comparison purpose, fivefold cross-validation approach, in which the training data D_1 was divided into 5 subsets of equal sizes [2], was also performed in this work. The comparison between fivefold cross-validation and holdout approaches are summarized in Table 6.9. The computational conditions were identical, namely, the population $m = 5$ and the maximum generations $G = 30$. The adjustable ranges of C and γ were [0.1, 500] and [0.001, 3], respectively. The fivefold

Table 6.9 Summary of the performance of ACO–SVR using two approaches

Approach	R		MRE (%)		Computing time (s)
	D_1	D_2	D_1	D_2	
Holdout	0.9629	0.9355	0.6020	1.5966	164
Fivefold	0.8868	0.8861	1.8222	2.0296	628

cross-validation method presented worse predictive accuracy and needed much more computing time than the holdout approach. Therefore, the holdout approach is suitable for this investigation in terms of the predictive accuracy and the computing time.

6.3.1.4 Prediction Results of Grid SVR

In the grid search method, two steps are needed in order to avoid the grid search with fine grid within a large area. First, optimal values of C and g are searched for in relatively coarse grids within a relatively large area, and then followed by a search in finer grids within a relatively small area. Finally, the optimal values of C and g for the model are obtained. Figure 6.13 shows the contour of MRE and R searched in the range of $C = (2^{-5}, 2^{-3}, \ldots, 2^{15})$ and $\gamma = C = (2^{-15}, 2^{-14}, \ldots, 2^{3})$. From these coarse contours of raw search, a smaller area was determined where a finer search $C = (2^{7}, 2^{7.25}, \ldots, 2^{9})$ and $\gamma = C = (2^{-10}, 2^{-9.75}, \ldots, 2^{3})$ was carried out to find the solution, as shown in Fig. 6.14. Finally, the lowest value of MRE and the highest value of R were found to be obtained at $C = 2^{7}$ and $\gamma = 2^{-0.75}$. After obtaining the optimal values of C and g with respect to the chosen grid point resolution, the model for NO_x emissions was then trained using these optimal parameters. The results are given in Table 6.10 in terms of MRE and R on the training set D_1 and the test data D_2. The total computing time for grid search was

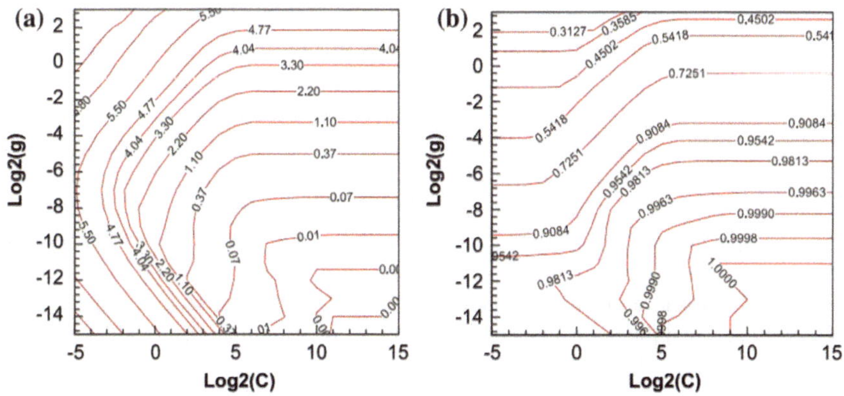

Fig. 6.13 Selection of (C, γ) by grid search on coarse grid: **a** MRE and **b** R

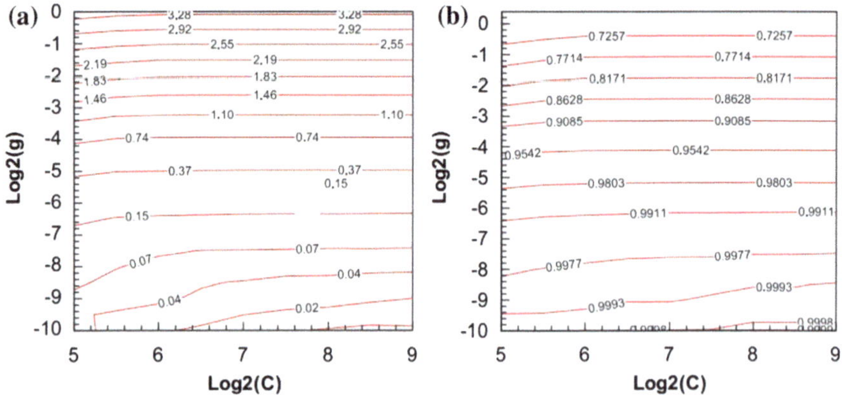

Fig. 6.14 Selection of (C, γ) by grid search on finer grid: **a** MRE and **b** R

Table 6.10 Summary of the performance of grid-SVR

	R		MRE (%)		(C, γ)	Computing time (s)
	D_1	D_2	D_1	D_2		
Coarse grid	0.9582	0.9355	0.7088	1.6092	(128.00,0.500)	193
Finer grid	0.9629	0.9360	0.6069	1.5986	(128.00,0.595)	195

about 388 s, the summation of the time spent on coarse grid search and that spent on finer grid search.

Figure 6.15 compares relative error distribution for the test data D_2 derived from two SVR models. The ACO–SVR model in Fig. 6.15 was obtained with a combination of the population $m = 5$ and the maximum generations $G = 30$. The horizontal axis represents the relative error ranges calculated by comparing the predicted NO_x emissions and the actual values, and the vertical axis represents the occurrence frequency of cases whose relative errors fall within the corresponding error range on the horizontal axis. The results in Fig. 6.11 show that the relative error distribution for ACO–SVR is almost identical to that of Grid-SVR. Judged by MRE, ACO–SVR shows slightly better results than Grid-SVR.

The main advantages of ACO–SVR over the Grid-SVR may be composed of two aspects. On the one hand, the ACO–SVR can automatically regulate the parameters pair (C, γ). On the contrary, Grid-SVR needs more manual operations. For example, two separate steps in Grid-SVR are not convenient to operate this process in automation. On the other hand, the computing time required for Grid-SVR was more than twice that of ACO–SVR (388 s vs. 164 s). Hence, it can be concluded that the performance of ACO–SVR is better than that of Grid-SVR for the current study. The ACO–SVR is considered as the “optimum” SVR model and will be compared to ANNs model in the next section.

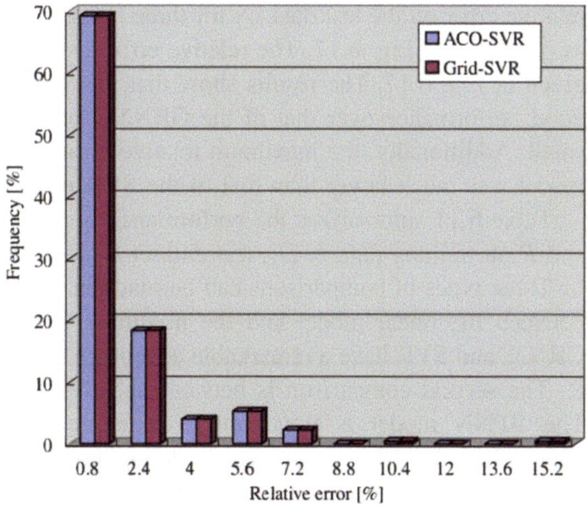

Fig. 6.15 Relative error distribution derived from two SVR models for NO_x emission

6.3.1.5 Comparison and Discussion

This section compared the predictive accuracy of the SVR model for NO_x emissions determined in the previous section to that of BPNN and GRNN models. As stated previously, the input data to BPNN and GRNN were the same as that for the SVR model. In addition, the holdout approach was also performed on BPNN and GRNN to improve generalization error. The evaluation criteria for each model were the MRE and the correlation coefficient *R*.

Figure 6.16 compared the predicted NO_x emissions derived from the "optimum" BPNN, the "optimum" GRNN and the "optimum" SVR. This figure is given as an illustration to provide insight into the performance of the models. As shown in Fig. 6.16, the blue dash-dotted line, by comparing with red-dashed line and the green dashed line, is obviously away from the black solid line. Hence, the predicted emissions by BPNN deviated from the measured emission more than the SVR and GRNN. Therefore, the predictive performance of GRNN and SVR was better than that of BPNN. From the observation of Fig. 6.16, one cannot distinguish the predictive accuracy of SVR model from that of GRNN model. The distribution of

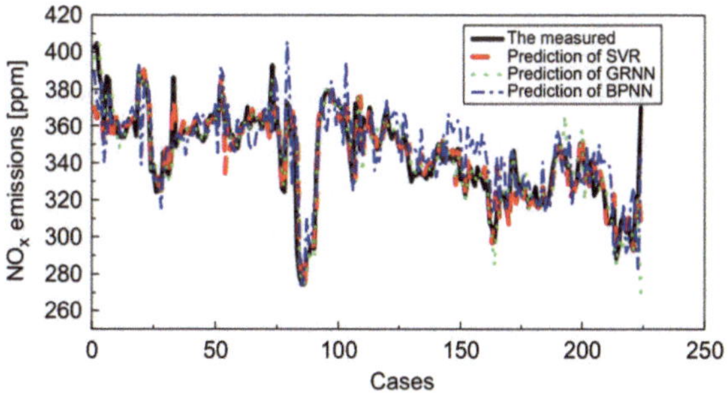

Fig. 6.16 Difference between estimation of the NO_x emission obtained respectively

relative error on the test data D_2 for three models was used for further comparison, as presented in Fig. 6.17. The relative error distribution derived from MLR is also given in Fig. 6.17. The results show that the SVR model demonstrated a slightly good performance over that of the GRNN model though the difference was very small. Additionally, the maximum relative error on the test data D_2 for the GRNN model was much larger than that of the SVR model, as shown in Fig. 6.17.

Table 6.11 summarizes the performance of various models. It shows the MRE and R on training dataset D_1, test dataset D_2, and the computing time.

Three types of comparisons can be made in Table 6.10. The first comparison is between the linear model and the nonlinear models. The latter such as BPNN, GRNN, and SVR have a remarkable advantage over the MLR in terms of MRE and R. The second comparison is between the BPNN model versus the SVR model. The BPNN model is important to examine because they are the traditional approaches for NO_x emissions modeling employed in PEMS and combustion optimization. Obviously, the SVR model performs significantly better than the BPNN model, as examined above.

The third comparison is made between GRNN and SVR. Both models are one-pass techniques, in which the certain results will be given once the model

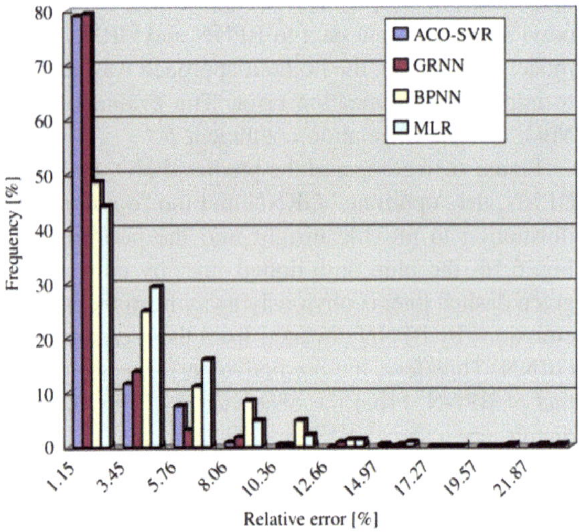

Fig. 6.17 Relative error distribution on the test set D_2 for various models

Table 6.11 Summary of various models for NO_x emissions modeling

Models	R		MRE (%)		Computing time (s)
	D_1	D_2	D_1	D_2	
MLR	0.7799	0.7729	3.3870	3.4390	<1
BPNN	0.9664	0.8890	1.3244	2.5060	4155
GRNN	0.9834	0.9267	0.5917	1.6217	981
ACO–SVR	0.9629	0.9355	0.6020	1.5966	164

parameters (smoothing parameter for GRNN model and C, γ for SVR model), unlike the BPNN model will give uncertain results even though the structure configuration, the number of the neurons in hidden layer and the training algorithms were determined in advance. The performance of the SVR was slightly better than the GRNN for NO_x emissions prediction in terms of MRE and R. In addition, as indicated by [3], some drawbacks restrict GRNNs wide application to various fields. On the one hand, the estimate from the GRNN is bounded by the minimum and maximum of the observations, namely, the predicted NO_x emissions for a new set of operational parameters of the boiler is always bounded between the maximum and minimum NO_x emissions of all cases because the estimated NO_x emissions can be visualized as a weighted average of all of the observations. Though this feature was regarded by [3] as one of the GRNNs advantages in relation to other nonlinear regression techniques, it is considered here as a disadvantage of GRNN from the combustion optimization's point of view. Taking the operational parameters optimization of the boiler, for example, if the GRNN model is incorporated into the optimization algorithms for optimizing the input parameter of the boiler, the optimized output will be not less than the minimum NO_x emissions in the training samples and the combustion optimization will lose its significance. So, searching tools such as GA and ACO are not suitable to be combined with the GRNN model for process optimization. On the other hand, another disadvantage of GRNN is that it requires substantial computation to evaluate new points.

To perform online and real-time application in PEMS and combustion optimization, the time response of modeling NO_x emissions is a critical consideration. Consequently, the comparison among algorithms is the calculation in real CPU time. The computing times are also compared in Table 6.10. A run of ACO–SVR optimization procedure can be completed in approximate 164 s, which is very suitable for online and real-time NO_x emissions modeling in coal-fired utility boilers. The process of finding the "optimum" GRNN and BPNN model took 981 s and 4522 s, respectively. The computing time required for GRNN model may be saved with the help of other optimization algorithms for finding the smoothing parameter. The main conclusion is that the SVR model has lower prediction error, shorter computing time, and a higher robustness than the BPNN model and GRNN model for NO_x emissions modeling.

6.3.1.6 Conclusions

In this part, different models for the estimation of NO_x emissions, obtained on the basis of data acquired on the output of an actual power plant, were introduced and their estimation capabilities were described. In particular, the performance (the predictive capability and the time response) of SVR model were compared to those of both BPNN and GRNN models.

Because of its universal approximation ability, SVR can be used to model nonlinear processes. Choosing optimal parameters for the models is an important step in the modeling stage. ACO optimization in combination with the holdout method is a reliable way to determine the optimal model parameters, referred to as ACO–SVR. Compared to the widely used grid search method for selection of parameters pair (C, γ), ACO–SVR needs less computing time and shows slightly better predictive accuracy. Additionally, the ACO–SVR can automatically select the optimal model parameters without any manual operation. The model obtained this way was found to have good generalization property. 92% of cases in test data D_2 had the relative error smaller than 5%. The predictive accuracy, characterized by MRE and correlation coefficient R, and the time response characteristics of the ACO–SVR model are greatly dependent on the product of the population and the maximum iterations G. The monotonic relations found between them can help us to make the trade-off. The optimization procedure with the population $m = 5$ and the maximum iterations $G = 30$ took 164 s to obtain a sufficiently accurate SVR model for NO_x emissions. The computing time in the scale of less than 3 min is suitable for online and real-time building of the SVR model.

Comparative study shows that the SVR model demonstrated much better performance than the BPNN model in terms of the predictive accuracy and the model robustness. SVR model also presented slightly better predictive accuracy than the GRNN model. The main advantage of the SVR model over the GRNN model is that the former can be incorporated into optimization algorithms such as ACO and GA to regulate the inputs of the model to achieve the expected NO_x emissions, which is the basic idea of combustion optimization to reduce NO_x emissions from coal-fired power plants.

In summary, the study demonstrated the capability of the artificial intelligence to model the complex and highly nonlinear relation of the NO_x emissions with the operational parameters of the boiler. The results of ACO–SVR can provide a guide to develop the PEMS for online NO_x emissions monitoring and combustion optimization software package for NO_x emissions reduction in coal-fired power plants. Chinese version: %%带有模型参数优化和运行参数寻优的PSO算法SVR案例%% English translation: PSO algorithm and SVR case in which both the model parameters and processing parameters are optimized.

6.3.2 *Modeling NO_x Emissions by SVM and PSO with Model and Operating Parameters Optimizing*

6.3.2.1 Experimental Setup

The experiments have been performed in dual furnaces tangentially fired full-scale boilers. A brief description of the test facilities is given below. Unit 8 of JianBi power station is a 300 MW power tangentially fired the boiler. The boiler is with a large furnace of 14.08 × 11.858 m^2 cross section and 53 m high. Six secondary air ports and four primary air ports are arranged on each corner of the furnace. Coal is pulverized by four medium-speed pulverizers with rotating classifiers. The burner arrangement is shown in Fig. 6.18.

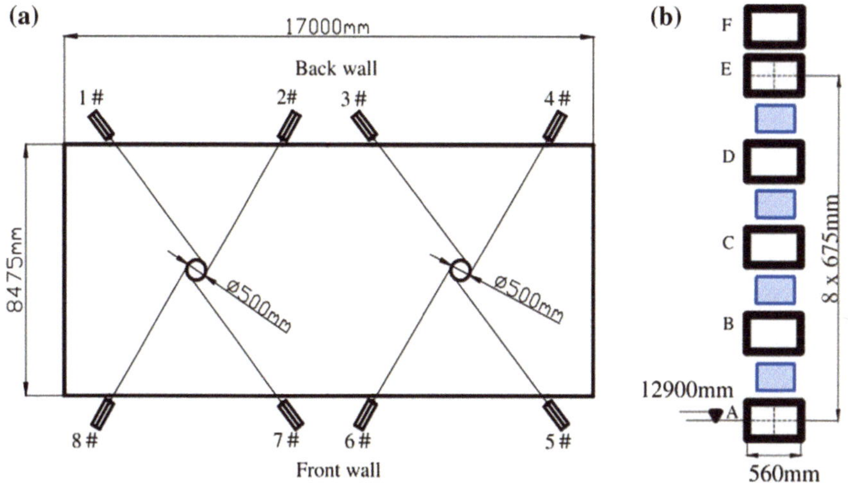

Fig. 6.18 Boiler schematic: **a** furnace cross-view and **b** burner arrangement

To obtain a wide range of operating conditions, a total of 670 tests have been carried out at this boiler to investigate the effects of the boiler load, primary air (PA), and secondary air (SA) distribution on NO_x emissions characteristics behavior. During the experiments, coal quality was kept constant, and NO_x concentrations were continuously monitored in the boiler outlet prior to the air heater by using continuous emission monitoring system (Rosemount, Emerson Process Ltd.). The measurements were performed a week later after the boiler switched to the pure test coal, to make the boiler conditions constant. The operating conditions during field tests are summarized in Table 6.12.

The NO_x emissions characteristics of this studied boiler were plotted as a function of the boiler load, as shown in Fig. 6.19. Each open cycle in Fig. 6.19 was responsible for a field test. Though scatter points existed in this graph, the NO_x emissions increased with the boiler load, as shown by the solid line (least squared fit between NO_x emissions and the boiler load). The correlation factor between the solid line and the scatter points is 0.2674. The positive correlation factor also meant increasing NO_x emissions with the increasing boiler load. However, the averaged deviation of the scatter points from the solid line is 19.44 ppm, thus leading to the insufficient prediction of NO_x emissions by the boiler load alone.

NO_x emissions characteristics model is probably the most important key to low NO_x combustion optimization in coal-fired power plants. SVR was also employed to predict NO_x emissions of the boiler in this work. Figure 6.20 presents the modeling error between the predicted and the measured NO_x emissions. The open circle presents the modeling error on the training subset, while solid inverse triangle represents the modeling error on the testing subset. The 224th case in the testing subset had the maximum modeling error of 16%. Also, 97% (647 cases) of the total 670 cases had modeling error less than 5%. The mean modeling error and the correlation factor were 1.08% and 0.95, respectively. As a whole, the predicted

Table 6.12 Operating conditions of the boiler during field tests

Operating conditions	Variation range
Primary air velocities (m/s)	
A level	[22.9, 29.3]
B level	[24.3, 29.2]
C level	[2.4, 28.3]
D level	[23.2, 29.1]
Secondary air velocities (m/s)	
A level	[24.7, 35.8]
B level	[23.8, 31.7]
C level	[23.4, 36.9]
D level	[20.9, 35.3]
E level	[19.7, 39.0]
F level	[2.1, 9.9]
Speed of pulverizers (rpm)	
A level	[497.7, 651.5]
B level	[237.2, 505.0]
C level	[0.0, 626.3]
D level	[300.9, 635.8]
Load (MW)	[239.6, 331.0]
Coal quality	
Volatile content (%)	31.96, constant
Moisture content (%)	11.06, constant
Ahs content (%)	13.6, constant
Low calorific value (MJ/kg)	23.49, constant
NO_x emissions (ppm)	[258.7, 407.2]

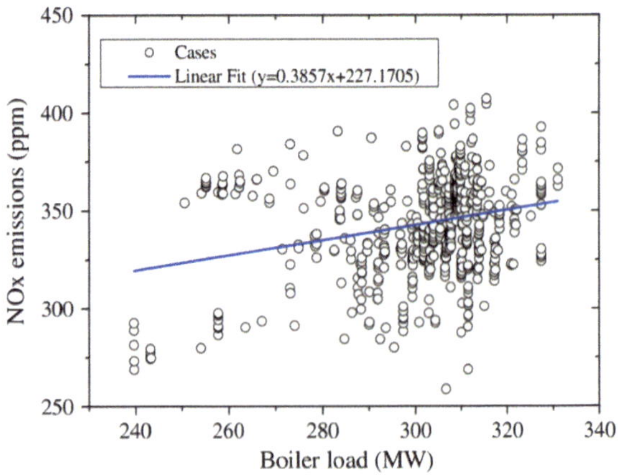

Fig. 6.19 Relationship between the boiler load and NO_x emissions

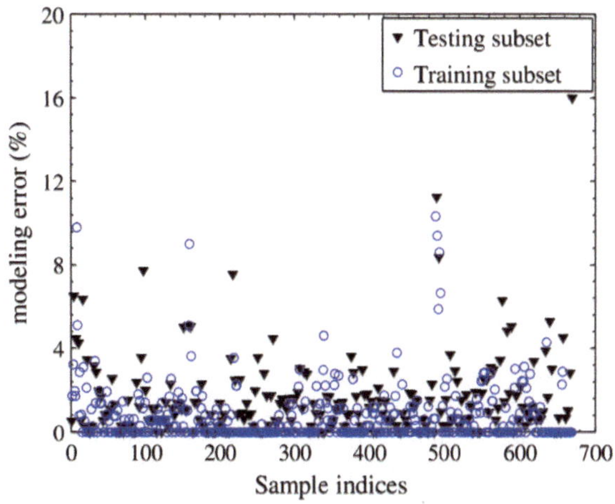

Fig. 6.20 Modeling error of the SVR model

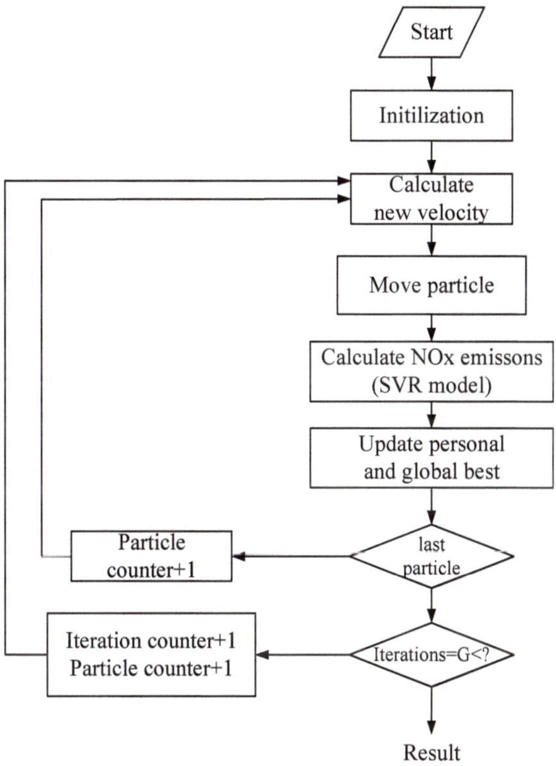

Fig. 6.21 Flowchart of the simulation

values showed rather good agreement with the measured values. The reliability and the accuracy of this SVR model have also demonstrated by in situ field tests.

The flowchart of the proposed method was given in Fig. 6.21.

6.3.2.2 Optimization Results for the Boiler Load of 288.45 MW

In each case study, 100 independent runs were made for each of the optimization methods. One main advantage of the PSO method is that the algorithm is relatively easy to tune, as the number of parameters to be adjusted is limited. The number of particles, *K*, is the only parameter to be tuned by users when the update rules of particles velocity by Clerc or by Trelea were adopted.

One case with boiler load of 288.45 MW and NO_x emissions of 318 ppm was chosen for optimization. The process iteration history was shown in Fig. 5. To compare the effects of update rules for particle velocity, parameters set by Clerc and Kennedy, by Trelea, and linearly decreasing weight with the initial weight of 0.9 and the end weight of 0.4 were mutually compared. Obviously, parameters set by Clerc and Kennedy outperforms those of Trelea and dynamic weight rules, as shown in Fig. 6.22a. The influence of the number of particles, *K*, was also investigated, as shown in Fig. 6.22b. It is clearly shown that the performance of the

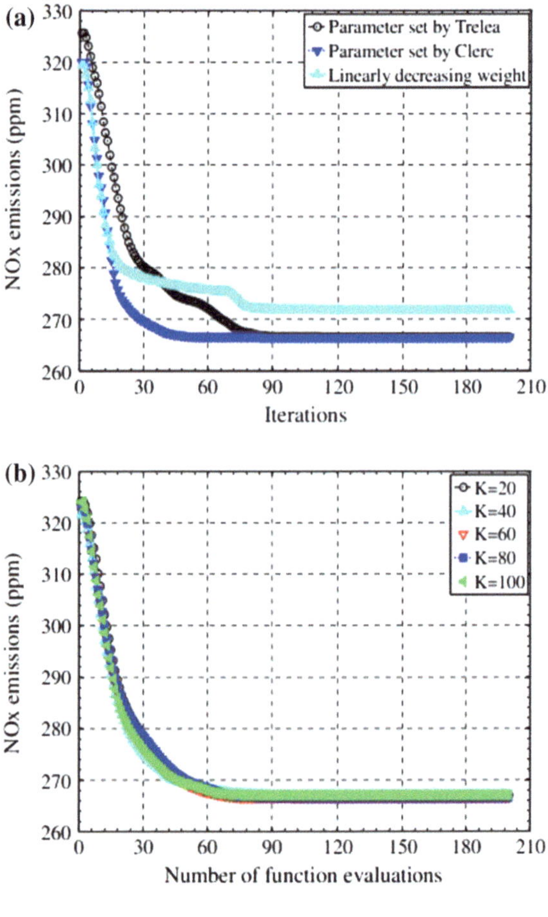

Fig. 6.22 Effect of: **a** update rule and **b** the number of particle on optimization results

proposed method was less dependent on the number of particles, which will help users to choose the appropriate parameters. On the whole, the proposed PSO method was a relatively robust and conveniently used optimization method.

6.3.2.3 Comparison with Other Methods

The proposed method was compared with the GA method by Zhou et al. and the ACO method by Zheng et al 22. To make the comparison fair, the control parameters for PSO, ACO, and GA were set carefully. In using the PSO algorithm, the population size, K, was set to 80, which was exactly the same as the population size of ACO and GA methods. The maximum iteration number is set to 500 ($G = 500$) for PSO, 1000 for both ACO and GA. With respect to the GA algorithm, the control parameters, i.e., crossover possibility and mutation possibility were obtained by the trial and error and set to $p_c = 0.625$ and $p_m = 0.275$. As well known, NO_x emissions concentration of the boiler is closely related to the load, as shown in Fig. 6.19. The cases with distinct loads probably give different NO_x emissions because many other operating conditions must be regulated according to the predetermined load. Generally, the boiler load is affected by the load dispatch scheme. That means the boiler load must be taken into account when considering NO_x emissions minimization. The aim of this work was to investigate the effectiveness of the present method on NO_x emissions minimization under low load and high load. Hence, we chose the two cases as the examples to check the effectiveness of the proposed method.

Case I has the maximum NO_x emissions concentration of 407.2 ppm among a total of 670 field tests. For a boiler load of 312.08 MW, the predicted NO_x emissions concentration indirectly calculated by the SVR model from the design variables as well as other operating conditions was 399.7 ppm. Hence, the performance of the SVR model for Case I showed good predictive accuracy with the modeling error of 1.98%. NO_x emissions concentration of Case II with the boiler load of 288.45 MW was 318.5 ppm. The predicted NO_x emissions concentration was 318.4 ppm with the modeling error of 0.03%.

Tables 6.13 and 6.14 list the optimal PA and SA settings (design variables), the optimization results (also standard deviation of repetitive simulation experiments) for Case I and Case II derived from PSO, ACO, and GA algorithms. Figure 6.23 showed the solution results. These results clearly demonstrate the superiority of the PSO algorithm to both the ACO and GA algorithms. The result obtained by this PSO corresponds to the optimum found in previous attempts, but the convergence is much faster.

Figure 6.23 was only employed to illustrate the convergence speeds of various methods in which the maximum iterations was solely chosen as the stopping condition. It is clearly shown that there existed a time waste in Fig. 6 resulted from the inappropriate stopping criteria (e.g., the algorithm will not be terminated until the maximum iterations G has been achieved). In the practical applications, a more flexible stopping criterion can be employed. In this work, a stopping criterion

Table 6.13 Operating conditions and optimization results of case I (load = 312.08 MW)

Operating conditions	Before optimization	PSO	ACO	GA
PA (m/s)				
A level	26.8	**27.7**	27.6	27.3
B level	26.7	**26.5**	26.6	26.3
C level	27.7	**25.4**	27.6	25.6
D level	25.9	**26.1**	26.0	25.2
SA (m/s)				
A level	34.4	**28.1**	28.1	28.3
B level	31	**29.3**	29.3	28.7
C level	33.8	**34**	34.2	34.4
D level	33.4	**32.2**	32.4	31.9
E level	32.5	**33.6**	33.4	33.8
F level	4	**1.6**	1.6	1.3
Predicted NO_x (ppm)	399.7	**269.3(5.8)**	269.4(5.7)	271.9(2.1)

Bold indicates the primary wind speed and the secondary wind speed obtained using the particle swarm optimization algorithm.

Table 6.14 Operating conditions and optimization results of case II (load = 288.45 MW)

Operating conditions	Before optimization	PSO	ACO	GA
PA (m/s)				
A level	27.4	**28.2**	28.3	28.7
B level	27.5	**26.6**	26.5	26.8
C level	26.4	**25.0**	25.1	26.0
D level	26.8	**27.2**	27.1	27.2
SA (m/s)				
A level	31.0	**28.0**	27.7	28.0
B level	28.2	**27.2**	27.3	26.3
C level	32.6	**31.0**	30.9	31.2
D level	30.8	**30.2**	29.9	29.5
E level	32.2	**32.6**	32.2	32.8
F level	3.0	**0.6**	0.65	0.35
Predicted NO_x (ppm)	318.1	**266.6(0)**	266.9(0.7)	271.3(3.2)

Bold indicates the primary wind speed and the secondary wind speed obtained using the particle swarm optimization algorithm.

consisting of expGoal = 250 ppm, no_change_in _iterations = 30, max_allowed_-time = 120 s and G = 500, as defined in Sect. 4.3, was used. The algorithm will be automatically terminated if and only if one of four stopping conditions was matched.

The convergence results for 100 runs are given in Fig. 6.24. For Case I, by comparing the base emissions (399.7 ppm), the mean optimization results derived from PSO, ACO, and GA methods were 269.3(5.9, standard derivation) ppm (32.67% reduction), 270.7(6.45) ppm (32.27% reduction) and 294.3(8.5) ppm (26.37% reduction), respectively. While for Case II with the base emissions of

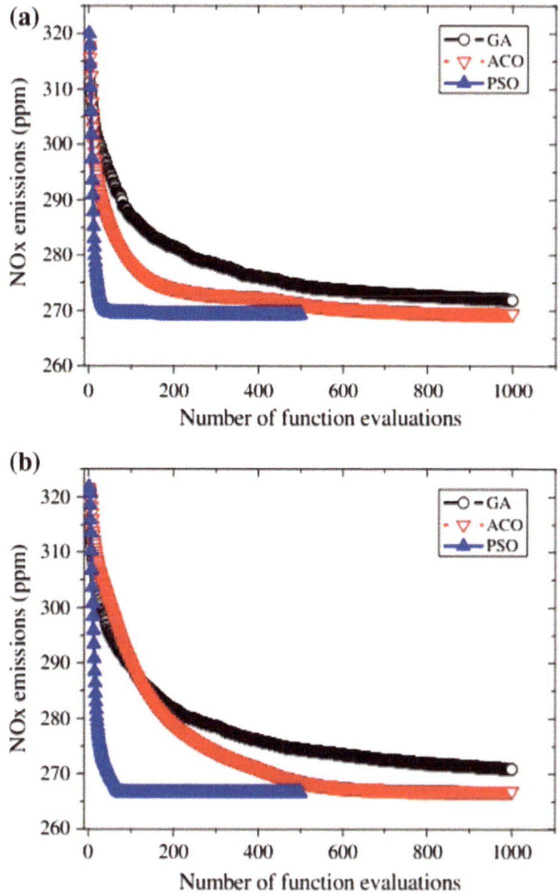

Fig. 6.23 Optimization processes of: **a** case I, load = 312.08 MW and **b** case II, load = 288.45 MW

318.4 ppm, the mean optimization results derived from PSO, ACO, and GA methods were 266.6(0) ppm (16.3% reduction), 270.9(7.9) ppm (14.92% reduction), and 284.9(7.9) ppm (10.5% reduction), respectively. Obviously, an improvement of solution quality was achieved by using the proposed method. By comparing Fig. 6.7; Tables 6.2 and 6.3, it is concluded that the optimized results were deteriorated for GA and showed nearly no change for PSO and ACO when the different stopping conditions were adopted. Taking the boiler load of 312.08 MW, for example, the averaged optimized NO_x emissions given by GA were 271.9 ppm for the simplex stopping criteria and were 294.3 ppm for the multiple stopping criteria.

By using the new stopping criterion, another advantage of using the PSO method is its computational efficiency. Computational time is monitored automatically (all the computational times shown were those averaged on 100 repetitive simulation experiments). A single CPU desktop computer (3.1 GHz, 1.5 GB RAM) was used. The optimal solutions (i.e., the objective function at the value of 269.3 ppm for Case I and of 266 ppm for Case II) were identified after 21.6 and 25.03 s,

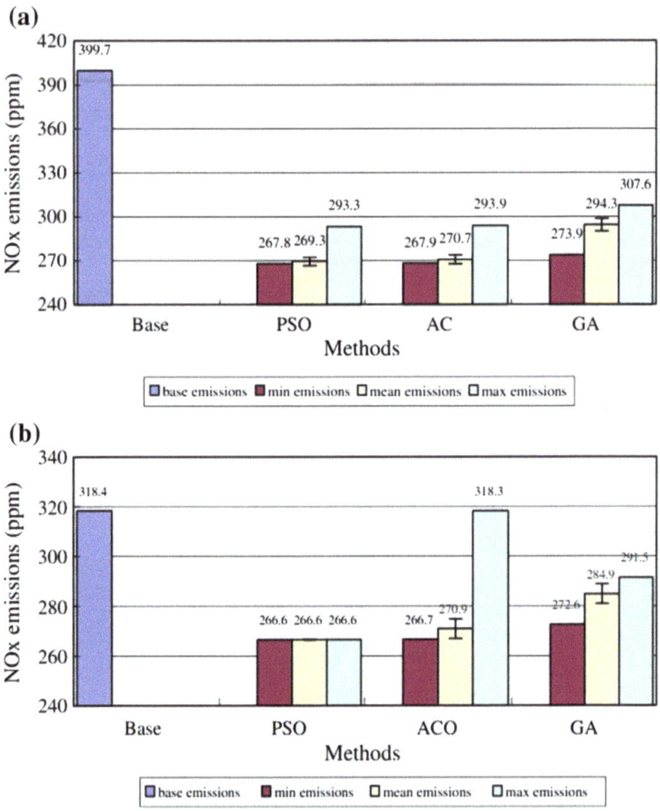

Fig. 6.24 Convergence results analysis of: **a** case I, load = 312.08 MW and **b** case II, load = 288.45 MW

respectively, as shown in Fig. 6.25. ACO proved more time demanding, requiring more than 120 s. A direct comparison of computational time makes it clear that the PSO method is very computationally efficient, which will be helpful for online and real-time applications. Moreover, the stopping conditions were more flexible. It is shown in simulation experiments that PSO and GA methods were terminated due to no_change_in_iterations = 30, i.e., no further improvement in solution within 30 successive iterations. However, GA presented poor solutions. With respect to ACO, the maximum allowed computational time becomes the main stopping condition.

6.3.2.4 Conclusions

In this work, the effects of operating parameters on NO_x emissions were experimentally investigated. The settings of operating conditions for coal-fired utility boiler must be studied carefully, as a set of inappropriate combustion parameters can lead to very high NO_x emissions. An empirical model calculating NO_x emissions from

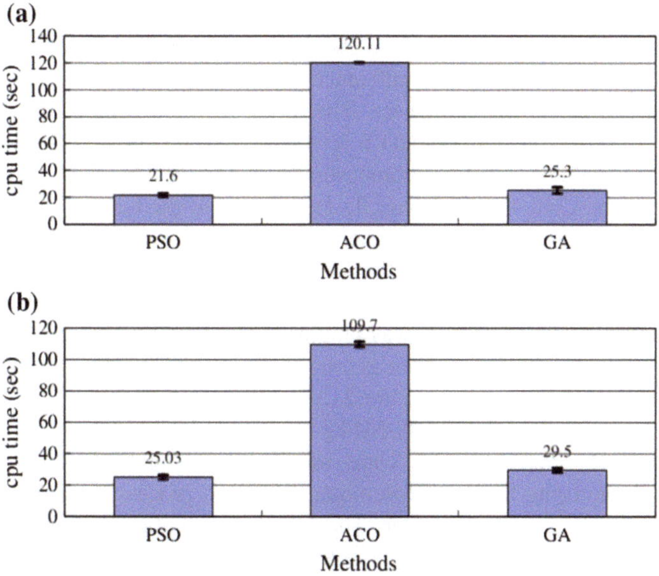

Fig. 6.25 Time responses of methods for: **a** case I, load = 312.08 MW and **b** case II, load = 288.45 MW

combustion parameters was represented by the SVR model, in which the 19 operating parameters of the boiler was chosen as the inputs of the model, the NO_x emissions as the output. A much simpler optimization tool, particle swarm optimization, was employed to regulate the inputs of the model, resulting in low NO_x emissions. The simulation experiments have shown that the proposed method was less influenced by the inherent control parameter, thus facilitating the design of optimization process. The performance of PSO was extensively studied by comparing its performance with GA and ACO methods. The result obtained by this PSO method reduced NO_x emissions by 32.67% at the load of 312.08 MW and by 16.3% at the load of 288.45 MW, respectively. With the typical computational cost of less than 25 s under a PC system, the running time of PSO was about one-fifth of those required for ACO.
Chinese version: %%各种寻优算法SVR比较案例%%
English translation: Comparison of SVR cases using different optimized algorithm.

6.3.3 *Comparison of Optimization Algorithms for Low NO_x Combustion*

6.3.3.1 Experimental Setup and NO_x Emission Data

Training data set were given from NO_x emissions of a power plant. The experiments were carried out in a 300 MW tangentially fired dry bottom boiler with large dual furnaces of 17 × 8.475 m^2 section and 45.5 m high, which is one unit of JianBi power plant located at Jiangsu Province of China. The fuel and primary air streams were directed at the circumference of two imaginary cylinders of 500 mm diameter at the center of the furnace. The tilting fuel and combustion air nozzles including 4 primary air burners and 6 secondary air burners were located in each corner of the furnace. Coal for combustion was supplied by four medium-speed coal pulverizers. Coal–air two-phase flows were fed to the burners on A–D levels. The boiler was manufactured by Shanghai Boiler Co. Ltd. in the 1980s. Operation experiences showed that NO_x emissions were high due to old-design technology of the boiler.

The concentric firing system was employed to combust bituminite. The characteristics of the combusted coal are as follows: the volatile content, the ash content, and moisture content as received are 31.96, 11.06, and 13.60 wt%, respectively. The heating value of coal is 23.49 MJ/kg.

NO_x and O_2 concentrations were monitored continuously by Continuous Emissions Monitoring System (Rosemount, Emerson process management) in the boiler outlet prior to the air heater. The NO_x concentrations reported here were average values over several hours of stable operation, and they were obtained under dry gas conditions. The measurements were performed a week later after the boiler switched to the pure test coal to make the boiler conditions unchanging. Six hundred and seventy tests were carried out on this boiler, changing the boiler load in the range of 239.64–331.03 MW, primary air velocity in the range of 2.39–29.28 m/s, A–E elevations of secondary air velocity in the range of 19.68–39.03 m/s, F level of secondary air velocity in the range of 2–10 m/s, respectively, to analyze the characteristics of NO_x emissions of the tangentially fired system. In all tests, coal quality remained constant. Due to the air duct arrangement restriction, only a very narrow air velocity change range of F level of secondary air can be achieved. The conducted experimental tests show that the boiler was operated in a wide range, the measured NO_x emissions and operational parameters provide a rich enough data set for developing an empirical model. The monitored NO_x emissions are shown in Fig. 6.26. The measured NO_x emissions varied from 258.7 to 407.22 ppm. Statistical analysis showed that more than 88% of cases in all tests emitted NO_x

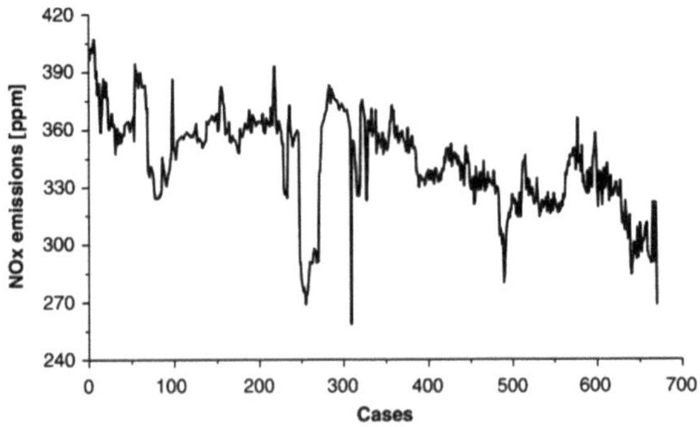

Fig. 6.26 NO_x emissions monitored for a coal-fired power plant

concentration higher than legislation requirement of China, 316 ppm. Methods must be taken to achieve low NO_x emissions so as to meet the requirement. The obtained operational conditions with 258.7 ppm NO_x emissions have demonstrated the potential of the boiler to achieve low NO_x emission by modifying operational parameters, which is also the basis we perform the current work. The database consisting of 670 cases with 22 dimensions was obtained, in which 21 inputs to the utility boiler were four levels of primary air velocities, six levels of secondary air velocities, oxygen concentration in the flue gas, speeds of four mills, the boiler load, total air flow rate and coal quality such as the ash content, the volatile content, the heating value, and the moisture content as received and one output was NO_x emissions concentration.

6.3.3.2 Estimation of NO_x Emissions by SVR

NO_x emissions prediction process by the SVR model is schematically shown in Fig. 6.27. Through physical analysis, it can be concluded that NO_x emissions in the coal-fired utility boiler were mainly determined by 21 operational parameters.

- Primary air velocity, including A, B, C, and D levels.
- Secondary air velocity, consisting of A, B, C, D, E, and F levels of secondary air velocity.
- The oxygen content in the flue gas.
- Rotation speeds of millers including A, B, C, and D.
- The boiler load.

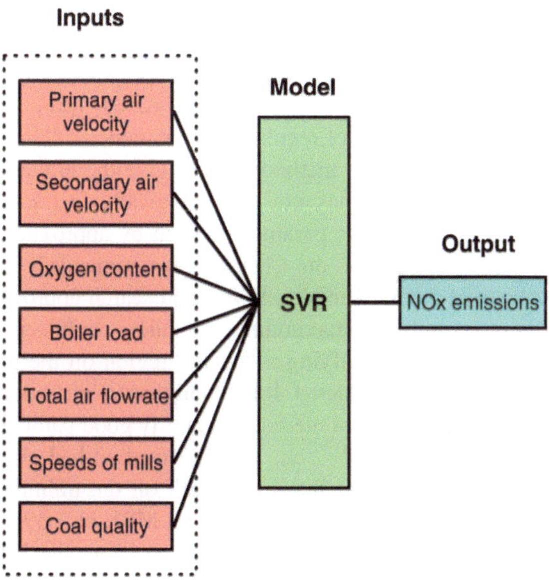

Fig. 6.27 Schematic of SVR model for NO_x emissions

- Total combustion air flow rate.
- Coal quality such as the volatile content, the ash content, the heating value, and the moisture content.

In the SVR model, these boiler parameters were used for inputs and NO_x emissions the output. After the training phase, the SVR model can be used to obtain the answer to a new input pattern.

Training the SVR model is an important step for developing a well-trained model. To do this, training data must be needed. Total 670 cases were classified into two parts, i.e., the training subset and the test subset. Cases were indexed by the Arabic numeral, i.e., $i = 1, 2, \ldots, 670$. Two hundred and twenty-four cases with the indices, $i = 1, 4, 7, \ldots, 670$ were chosen as the test subset, while remaining 446 cases were used to train SVR to capture the quantitative relation between NO_x emissions characteristics and 21 operational parameters of the studied utility boiler.

6.3.3.3 Selection of Model Parameters

There exists a problem in the practical application of SVR. This problem is how to select some SVR parameters so that the performance of SVR can be brought into the best. To date, there is not a very effective method to select these SVR parameters. The trial and error process is commonly used to tune up them. In fact, selection of SVR's parameters is an optimization issue. Based on this idea, GA was employed to determine the optimal parameters of SVR [4, 5]. The method was tried to build the SVR model for NO_x emissions. It took nearly 4 h to get the best parameters with the minimum MRE, the optimal pairs of (C, γ) was (311.50, 0.58). Training a SVR model from the dataset consisting of 670 samples with 22 dimensions may typically take several seconds. On average, GA with 50 initial populations requires typical 100 iterations to converge. Hence, at least 5000 SVR models are tried. The searching process is extremely CPU demanding.

Recently, Cherkassky and Ma (2004) proposed an empirical method for selecting the value of regularization parameter C for SVR directly from the training data. However, the method did not give the empirical expression for the value of c. Moreover, we may not psychologically feel safe to use methods which avoid doing an exhaustive parameter search by approximations or heuristics.

A "grid search" on C and γ, recommended by [6, 7], was widely used to construct the SVR model. In the grid search approach, pairs of (C, γ) are tried and the one with the maximum correlation coefficient R or the minimum MRE is picked. After identifying a "better" region on the grid, a finer grid search on that region can be conducted. It is found that trying exponentially growing sequences of C and γ is a practical method to identify good parameters (for example, $C = 2^{-5}, 2^{1}, \ldots, 2^{15}$, $\gamma = 2^{-15}, 2^{-9}, \ldots, 2^{3}$). The method had also been used here.

After conducting the grid search on the training data, the optimal (C, γ) was (362.0387, 0.3536) with a minimum MRE of 1.6902% for test subset. Figure 6.28 summarized results of the grid search using the MRE as an evaluation criterion.

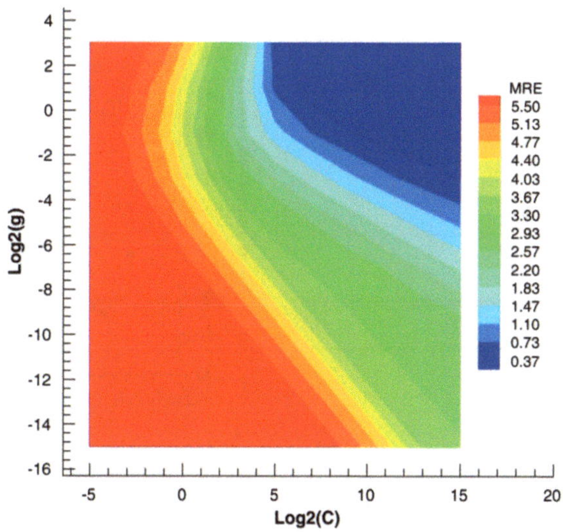

Fig. 6.28 The selection of (*C*, *c*) by grid search

As shown in Fig. 6.28, parameters pair fell in the top right area of the figure give smaller MRE. As the γ was fixed, it is found that the larger the value of *C*, the smaller was the MRE. It is an intuition that the large *C* is a good choice. However, caution must be taken that the SVR model will be insensitive to the new input when the large *C* is applied. For example, SVR produced almost the same NO_x emissions for two sets of operational parameters of the boiler, even though these two sets of operation parameters deviated significantly from each other.

The advantage of the grid search is that the time demanding for finding the optimal (*C*, γ) is acceptable. For example, the computing time to build the optimal SVR model for NO_x emissions by using grid search is in scale of minutes (approximately 8 min), which is much less than that required by GA search method. The method is also suitable in real time establish the SVR model.

6.3.3.4 NO_x Emissions Prediction Results

Figure 6.29 gives the comparison between the predicted NO_x emissions and the measured NO_x emissions. Solid line represented the perfect line with the predicted NO_x emissions equal to the measured. The blue solid circles represented the predicted NO_x emissions for training cases, while the solid triangles represented the predicted NO_x emissions for the test cases. The predicted NO_x emissions in Fig. 6.29 were in good agreement with the measured NO_x emissions as a whole. The 6th case (the definition of the index is given above) had the maximum NO_x emissions, 407.22 ppm. The predicted NO_x emissions of the SVR model for this case was 405.732 ppm, so the absolute error was very small. The absolute error and the relative error are very easy to be calculated for all cases from the figure. The

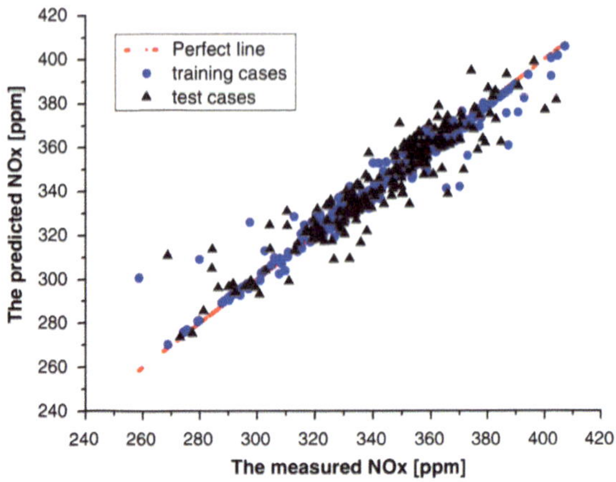

Fig. 6.29 Measured and predicted NO_x emission by SVR

maximum absolute error was about 42 ppm, given by 309th case, while the maximum relative error was 16.29%, also given by the 309th case. The case was the training sample. The generalization performance of the SVR model is better in contrast to that of the BPNN, in which the maximum error is often given by the test cases other than the training cases. With the exception of the 309th case, the relative error for most of the cases is below 5%.

6.3.3.5 Low NO_x Emissions by Combining SVR and Optimization Methods

Another benefit resulting from the SVR model is that it can be combined with optimizing algorithms. In general, various algorithms share a common approach for their application to a given problem. The general optimization process for low NO_x combustion is shown in Fig. 6.30. Prior to application of optimization, the objective function must be defined at first. In this study, the purpose is to reduce NO_x emissions by modifying the operational parameters of the boiler. Therefore, the objective function should be NO_x emissions, while the optimized variables should be operational parameters. If taking NO_x emissions function expressed by SVR as the fitness function, then the optimum operating parameters can be achieved using various optimization methods. As illustrated in Fig. 6.27, the SVR model for NO_x emissions has 21 inputs. However, not all inputs of NO_x emissions SVR model can be considered as optimized variables. Under a determined unit load, the most concerned parameters are the flue gas oxygen and the secondary air velocity and the primary air velocity. The boiler load and coal properties may be determined by the unit operating conditions and not be considered as adjustable parameters. The other parameters such as mill feeder speeds, total air flow rate may be determined by the boiler load, they are usually not be adjusted to decrease NO_x emissions.

Fig. 6.30 Low NO_x combustion optimization process

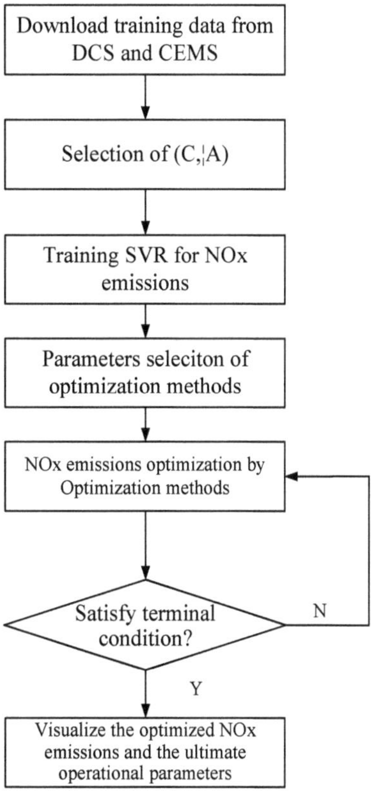

Through the above analysis, we chose four primary air velocities and six secondary air velocities and the oxygen concentration in the flue gas, as the parameters to be optimized. Therefore, the present optimization issue consists of one objective and eleven variables to be optimized. Of course, the other parameters can also be optimized with the same method from the optimization algorithms itself point of view. The adjustable range for the primary air velocity was 25–30 m/s, which is constrained by pulverized coal transportation. The adjustable ranges for the A–E levels of secondary air velocity, F levels of secondary air velocity, and oxygen content in the flue gas were 25–45 m/s, 0–25 m/s, and 2–5%, respectively, based on the operating routine and security considerations, which mean that the search space for the optimization methods was determined. The adjustable ranges for these operating parameters can be easily regulated if needed.

The purpose is to develop a fast and effective optimization algorithm to reduce NO_x emissions. The best optimization algorithm with the quick convergence rate and good performance is highly preferred with respect to the performance and time response requirements imposed by online combustion optimization. This subsection presents a comparison among four optimization algorithms in terms of performance, convergence rate. The results will give some insight into the selection of the appropriate algorithms to achieve low NO_x combustion modification.

6.3.3.6 Parameter Settings for Various Algorithms

Each algorithm has its own parameters that affect its performance in terms of solution quality and the computing time. Tuning the parameters for any optimization algorithm is at least as important as designing the algorithm itself [8]. To obtain the most suitable parameter values that provide the best performance for the test problems, a repetitive trial and error procedure was carried out. For GA, three combinations of parameters were used: (1) a crossover probability $p_c = 0.8$ and a mutation probability $p_m = 0.15$; (2) a crossover probability $p_c = 0.5$ and a mutation probability $p_m = 0.1$; (3) a crossover probability $p_c = 0.2$ and a mutation probability $p_m = 0.05$. The results showed that a combination of $p_c = 0.5$ and $p_m = 0.1$ led to lowest NO_x emissions. Upon experimentation, the suitable parameters for PSO were as follows: acceleration coefficients c_1 and c_2 were both equal to 2 and remained unchanged in the searching process; the lower and upper bounds of the inertia weight factor ω were 0.4 and 0.9, respectively. The inertia weight factor x decreased linearly with the iteration process. The controllable parameters which affect the performance of ACO[1] include the initial pheromone τ_0, the exploitation probability factor q_0, the pheromone decay parameter ρ and the local pheromone update strength parameter α. We checked the influences of m, τ_0, ρ, and α on the ultimate NO_x emissions. Results showed that pheromone decay parameter ρ and local pheromone update strength parameter α had little effects on the ultimate solution, while the quality of the solution improved with the increased number of ants m and the initial pheromone τ_0. However, these two parameters did not affect the ultimate solution when the number of ants m was larger than 20 and τ_0 was larger than 0.01. The trial and error process helped us to choose $\alpha = 0.1$, $\rho = 0.2$, $q_0 = 0.5$, $\tau_0 = 0.01$ as the parameters of ACO[1] algorithm afterward. As for ACO[2], after extensive experimentation, the parameters were set as follows: traverse step of other ants λ was a random quantity in the range of [0.1, 0.14]; $p_0 = 0.23$; the evaporation factor of pheromone $\rho = 0.9$; w_{max} and w_{min}, were random quantities in the range of [1, 1.4] and [0.2, 0.8], respectively; the traverse step of the iteration best ant w decreased linearly with the iterations.

6.3.3.7 Performance Comparisons

In order to ensure a reasonably fair comparison, we allowed for each application of four algorithms the same number of population size $m = 80$ (a population of 80 ants for two ACOs, 80 particles for PSO and 80 individuals for GA) and the maximum number of iteration $G = 500$.

To conveniently compare performance (or the quality of results) of four optimization algorithms, they were, respectively, applied to a special case. The selected case was the 6th one because it had the maximum NO_x emissions, 407.22 ppm. The task of reducing NO_x emissions for this case is very troublesome. The progressive processes of four algorithms for this case are presented in Fig. 6.31. The black squares, the red circles, the green triangles, and the blue inverse triangles,

Fig. 6.31 Comparison among four algorithms

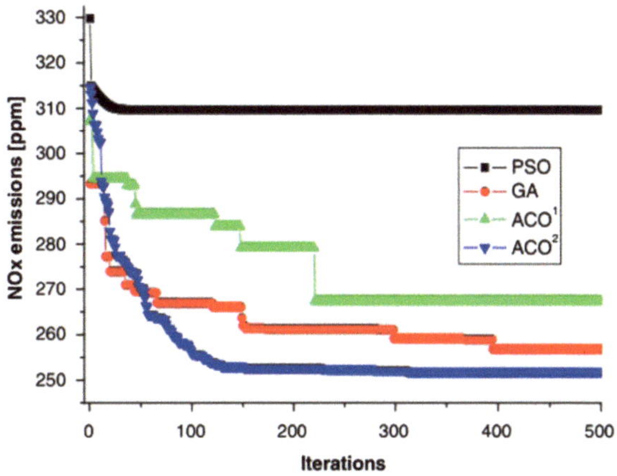

Fig. 6.32 Comparison between the original and the optimized NO_x emissions for the 6th case by various optimization algorithms

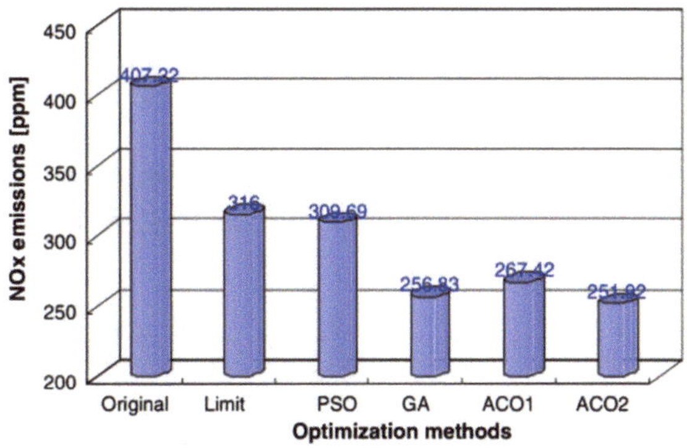

respectively, connected the optimized NO_x emissions under various iterations derived from the PSO, GA, ACO^1, and ACO^2. Comparing the original NO_x emissions and the optimized NO_x emissions is a straightforward measure of evaluating the performance of various optimization algorithms, as presented in Fig. 6.32. The original NO_x emissions for this case before optimization and the legislation limit of China were also given in Fig. 8. The original NO_x emissions was about 407.22 ppm for this case, while the optimized NO_x emissions by PSO, GA, ACO^1, and ACO^2 were 300.45, 256.83, 267.42, and 251.92 ppm, respectively. For this case, four algorithms all decreased the NO_x emissions below the limit, 316 ppm. Judged by the algorithms performance, the order of four algorithms by ascending sort was PSO, ACO^1, GA, and ACO^2 (see Fig. 8). That means PSO had the worst performance, while ACO^2 outperformed other three algorithms.

Operational parameters of the boiler for this case before and after optimization are listed in Table 6.15. The oxygen content decreased after optimization. Theoretical analysis shows that the decrease in oxygen content in the flue gas will

Table 6.15 Operational parameters of the boiler for the 6th case before and after executing optimization

Optimization variables	Primary air velocity (m/s)				Secondary air velocity (m/s)						Oxygen content (%)
	A	B	C	D	A	B	C	D	E	F	
Before optimization	26.81	26.74	27.69	25.94	34.43	30.99	33.84	33.36	32.53	4.8	2.61
PSO	28.91	25.51	25.05	26.14	26.6	31.22	29.05	33.16	33.41	3	2.62
GA	27.5	25.68	28.88	25	27	29.27	33.22	32.6	37.6	1.68	2.56
ACO1	26.9	26.3	25.7	25	28.02	29.88	29	31.82	36.5	2.51	2.32
ACO2	27.75	25.52	25.85	24.83	27.66	30.3	32.64	31.18	35.98	1.04	2.53

lead to the decrease in NO_x emissions. The optimized operational parameters, especially the secondary air velocity distribution among various levels, differed markedly from the original. The optimized A–E levels of secondary velocity distribution was inverse tower-like or inverse triangle-like, which means the velocities or air amount decrease gradually from the upper level to lower level of the burners group. Inverse tower-like secondary air distribution supplies less amount of the combustion air in the main combustion zone at the lower zone of the furnace to provide a fuel rich environment (high fuel/air stoichiometric ratio), as a result, leading to lower furnace peak temperature and less NO_x emissions. The oxygen-rich and fuel-lean condition resulted from more amount of combustion air at the upper zone of the furnace is helpful for the coal burnout. In general, the larger the height of the burners group, the more NO_x emissions reduction can be achieved by regulating the configuration of secondary air velocity. With an exception, F level of secondary air contributed little to low NO_x emissions due to very low velocity. The theoretical analysis shows that the optimized operational parameter listed in Table 6.15 are consistent with the physical nature of achieving low NO_x emission in coal-fired utility boilers by using low oxygen content combustion and rearrangement of secondary air distribution, and that the optimized results are reasonable.

The optimized NO_x emissions for all 670 cases by various optimization algorithms are shown in Fig. 6.33. NO_x emissions before and after application of SVR and various optimization technologies were plotted as a function of cases. All optimization algorithms except PSO can decrease NO_x emissions of all cases below the limit. The encouraging results show that NO_x emissions in coal-fired power plants can to a certain extent meet the environmental standard by only modifying operational parameters of the boiler instead of installing expensive flue gas treatment equipment such as SCR and SNCR. It was also demonstrated the combustion optimization based on artificial intelligence such as SVR and evolutionary-based algorithms provided a cost effective and promising way to reduce NO_x emissions from coal-fired utility boilers. However, in all optimization approaches, PSO can

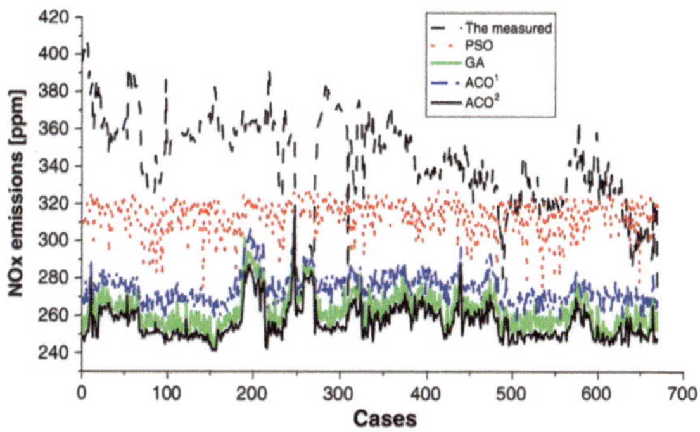

Fig. 6.33 Comparison of NO_x emissions for all cases before and after optimization

only decrease NO_x emissions of some special cases below the limit. Therefore, PSO implemented in this work cannot achieve deep NO_x emissions reduction and is not suitable for application in combustion optimization.

It is very interesting to compare ACO^2 to GA, because GA is a widely used optimization method for combustion modification. To examine the performance of ACO and GA in detail, the reduction percentage of NO_x emissions after optimization was regarded as an evaluation measure, which was defined as follows:

$$\text{Reduction Percentage} = \frac{\text{NO}x|_{\text{measured}} - \text{NO}x|_{\text{optimized}}}{\text{NO}x|_{\text{measured}}} \times 100\%,$$

where $NO_x|_{measured}$ is the originally measured NO_x emissions, $NO_x|_{optimized}$ is the optimized NO_x emissions. The reduction percentages of all cases for GA and ACO2 are shown in Fig. 6.34. The reduction percentage for GA varied from −3.9 to 37.6%, while that for ACO^2 was in the range of [0.42, 38.9%]. The results in this work are comparable to the NO_x reduction of over 30 percent reported at the Ameren Labadie station [9]. Negative values for the reduction percentage for case 258 and case 309 mean that the optimized NO_x emissions were slightly larger than the measured NO_x emissions. For the two cases, compared to GA, ACO^2 had also given only two minor reduction percentage −0.59 and 0.42%, respectively. The cause may be that these cases had already achieved low NO_x emissions close to the optimum. Results in Fig. 10 present an obvious fact that the performance of ACO^2 outperformed that of GA because the reduction percentage for the former was always higher than that for the latter. Socha [10] and Socha and Blum [11] have reported that a direct comparison between ACO and GA demonstrated the superiority of ACO over GA in benchmark function optimization [10] and weights and bias value optimization of back propagation neural network [11]. Results presented here probably confirm for the first time that the performance of the well-designed ACO is better than that of GA in engineering application. It is important to note that for those cases with relatively low NO_x emissions, their emissions cannot be further decreased only by optimizing the primary air velocity, secondary air velocity and

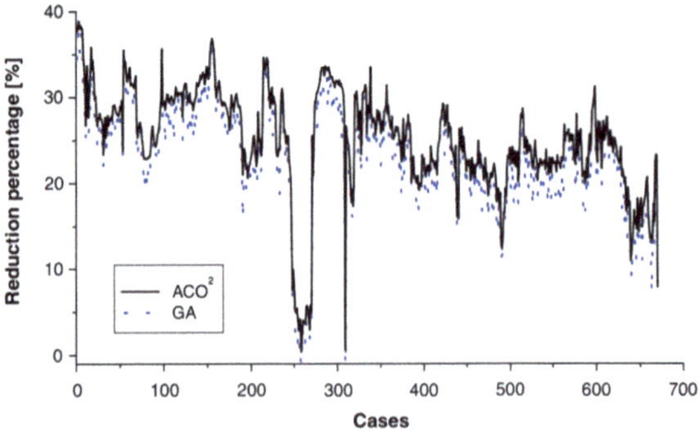

Fig. 6.34 NO_x reduction percentage by GA and ACO^2

oxygen content even though the well-designed optimization algorithms (for ACO^2 instance) were applied. For example, NO_x emissions for case 309 were 258.7 ppm, while the optimized NO_x emissions were only 257.18 ppm. The mechanism of the NO_x formation in coal-fired utility boiler shows that the combustion modification can only partially reduce NO_x formation to a certain level. Further deep NO_x emissions reduction must be only achieved by other physical technologies such as flue gas treatment methods, rather than combustion modification. Therefore, the results derived from optimization algorithms agree with the physical aspect of the NO_x formation in combustion process.

6.3.3.8 Convergence Rate

Time response characteristic of optimization algorithms is another key consideration in practical application. The quick speed of finding the optimal operational parameters is highly desirable for online and real-time combustion modification. As shown in Fig. 6.31, judged by the convergence rate, PSO is the fastest one. Nevertheless, the quick speed of the algorithm becomes spurious, so that it will give poor results in actual application. In fact, PSO had been trapped into the local minimum very fast and showed the worst performance among four optimization algorithms. ACO^1 took about 250 iterations to obtain the ultimate NO_x emissions, while ACO^2 took less than 150 iterations to achieve a steady solution. Compared to ACO1 and ACO^2, GA showed a much slower progressive process. In the previous study [12], GA needed to take about 600 iterations to find the minimum NO_x emissions. As also indicated by [13], GA requires long processing time for a near optimum solution to evolve. The iterations taken by various algorithms do not mean the actual time they spent because the time spent on each generation is different from one algorithm to another. It is more straightforward to evaluate the time response by comparing the CPU time or wall-clock time of various algorithms. The computing times of various optimization algorithms were presented in Fig. 6.35.

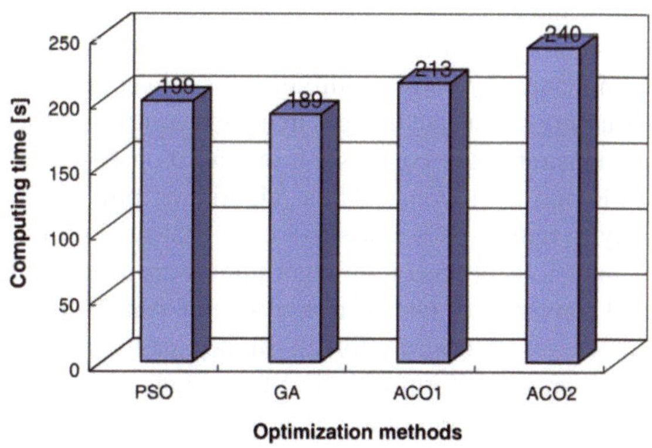

Fig. 6.35 Comparison between computing times of various optimization algorithms

Results in Fig. 6.35 were obtained on a desktop computer with 1.5 G DRM and 2.4 GHz CPU. The computing time of ACO^2 algorithm was the most among all algorithms. Although the computing time spent on a run of optimization procedure for other algorithms was less than that for ACO^2, ACO^2 took smaller iterations to achieve a steady solution. Therefore, the absolute time required for a run of ACO^2 can be saved by setting smaller the maximum iterations G. It is easy to calculate that the wall-clock time required for ACO^2 will be less than that of other optimization approaches by reducing the maximum iterations without any loss of the ultimate performance.

In summary, it is concluded that ACO^2 optimization method, for cases studied in this work, showed better results than the other optimization approaches in terms of the performance (the ultimate NO_x emissions) and the convergence rate. The computing time to execute a run of optimization procedure was in the scale of 4 min, which is acceptable for the online and real-time application to actual power plants.

6.3.3.9 Conclusions

The goal of this study is to compare various optimization algorithms applied in combustion modification to reduce NO_x emissions from a power plant. Modeling and optimizing NO_x emissions are two important steps in combustion modification. Therefore, the first stage was to model the functional relation between NO_x emissions and operational parameters of a utility boiler by using support vector regression. The predicted NO_x emissions from the SVR model showed good agreement with the measured. Compared to genetic algorithm-based method, the grid search method on selection of two parameters (C, γ) is suitable for on-line establishment of the optimum NO_x emissions SVR model. To realize low NO_x combustion modification by carefully setting operational parameters of the boiler, four optimization algorithms were, respectively, presented. The results show that the hybrid algorithm by combining SVR and optimization algorithms (with the exception of PSO) can effectively reduce NO_x emissions of coal-fired utility boilers below the legislation requirement of China. The encouraging results show that NO_x emissions in coal-fired power plants can to a certain extent meet the environmental compliance by only modifying operational parameters of the boiler instead of installing expensive flue gas treatment equipment such as SCR and SNCR. It is also demonstrated the combustion optimization based on artificial intelligence such as SVR and evolutionary-based algorithms provides a cost-effective and promising way to reduce NO_x emissions from coal-fired utility boilers. Detailed comparisons among four optimization algorithms were made in terms of the quality of results and convergence rate. The results show that PSO have the worst quality of solution among all algorithms for all studied cases. The well-designed ant colony optimization outperforms classical genetic algorithm. The presented study is favorable to the development of the sophisticated combustion optimization software package.

6.4 Multi-objective Optimization of Coal Combustion for Utility Boilers

6.4.1 Multi-objective Optimization Algorithm

A multi-objective optimization problem can be formulated as

$$\begin{cases} \min f(\vec{x}) = [f_1(\vec{x}), f_2(\vec{x}) \ldots, f_n(\vec{x})] \\ \text{s.t. } g(\vec{x}) \leq 0 \\ h(\vec{x}) = 0 \end{cases} \tag{6.25}$$

where $\vec{x} = [\vec{x}_1, \vec{x}_2, \ldots, \vec{x}_m]^{\mathrm{T}}$ is a vector of decision variables in the decision space X. f_i is an objective function, i = 1, 2, …, n, while $h(\vec{x})$ and $g(\vec{x})$ are the inequality and equality constraint vectors. The multi-objective problem is defined to find the solution set in the decision space that will satisfy all constraints and yields the optimal values to the objective functions. In the absence of any further information, one of these optimal solutions cannot be said to be better than the other.

$$\begin{cases} \forall 1 \leq i \leq n f_i(\vec{x}_1) \leq f_i(\vec{x}_2) \\ \forall 1 \leq j \leq n f_j(\vec{x}_1) < f_j(\vec{x}_2) \end{cases} \tag{6.26}$$

6.4.1.1 The Cellular Genetic Algorithm for Multi-objective Optimization (MOCell)

It has been proved that the cellular models of genetic algorithms (cGAs) are very effective in solving a diverse set of single objective optimization problems from both real-world and classical settings (Alba and Dorronsoro 2005). The cellular genetic algorithm for multi-objective optimization (MOCell) is proposed by Nebro et al. [14], which is an adaptation of a canonical cGA to the multi-objective field. The additional feature of MOCell is to return a number of solutions, from the archive to the population by randomly replacing existing individuals [14]. This makes MOCell distinct from other multi-objective evolutionary algorithms. In MOCell, the individual is distributed according to a certain topology and only interacts with its nearby neighbors in the breeding loop. The most commonly used population topology is two-dimensional toroidal grids, which is adopted in this chapter. All the individuals distribute in the grids, and each individual occupies a certain grid, and its neighbors including eight individuals adjacent with it in the east, south, west, north, southeast, southwest, northeast, and northwest directions. The detailed process of MOCell may found in the study by Nebro et al. [14], it is shown as follow.

```
Proc Steps UP(mocell)
Pareto_front = Create_Front()                              //Creat an empty
Pareto front
   While !TerminationCondition() do                        //Until terminal condition is satisfied
      For individual ← 1 to mocell.popsize   do            //Do
         n_list← Get_Neighborhood(mocell, position(individual))
        parents← Selection (n_list);             // Select parent population from neighbour
Senoffspring←Recombination(mocell.Pc, Parents) //Crossover to offspring population
        offspring←Mutation(mocell.Pm,offspring);     //Mutating on offspring population
           Evalutate_Fitness(offspring);               //Estimate offspring population
     Insert(position(individual),offspring,mocell,aux_pop); //Insert individual into auxiliary one
      Insert_Pareto_Front(individual);     //Insert individual into Pareto front
      End For
      Mocell.pop←aux_pop;                  //Replace original population with auxiliary one
      Mocell.pop←Feedback(mocell,ParetoFront);                //Do feedback
   End while
End_proc Dteps_up
```

6.4.1.2 AbYSS Algorithm

Scatter search has been successfully applied to a wide variety of single objective optimization problems [15]. AbYSS is proposed by Nebro et al. [14], which is based on the well-known scatter search template [16]. However, the AbYSS cannot be considered as scatter search but a hybrid of this algorithm with an evolutionary algorithm, because AbYSS combines ideas of state-of-the art evolutionary algorithms for solving multi-objective problems. In this algorithm, an external archive is used to store the non-dominated solutions founded during the search, following the scheme applied by PAES, however, the adaptive grid used by PAES is instead by using the crowding distance of NSGA-II; on the other hand, the density estimation used by SPEA2 is used to the selection of solutions from the initial set to build the reference set. The detailed process of AbYSS could be found in the study by Nebro et al. [14]. The flow chat is shown as Fig. 6.36.

6.4.1.3 OMOPSO Algorithm

The swarm strategy for optimization is first introduced by [17] and the particle swarm optimization (PSO) algorithm is based on the simulation of the social behavior of birds within a flock. The PSO algorithm has been applied in many single-objective optimization problems. To handle multi-objective problems, however, PSO algorithm must be obviously modified.

In the OMOPSO algorithm, a crowding factor [18] is used to establish a second discrimination criterion (additional to Pareto dominance). For each particle, the

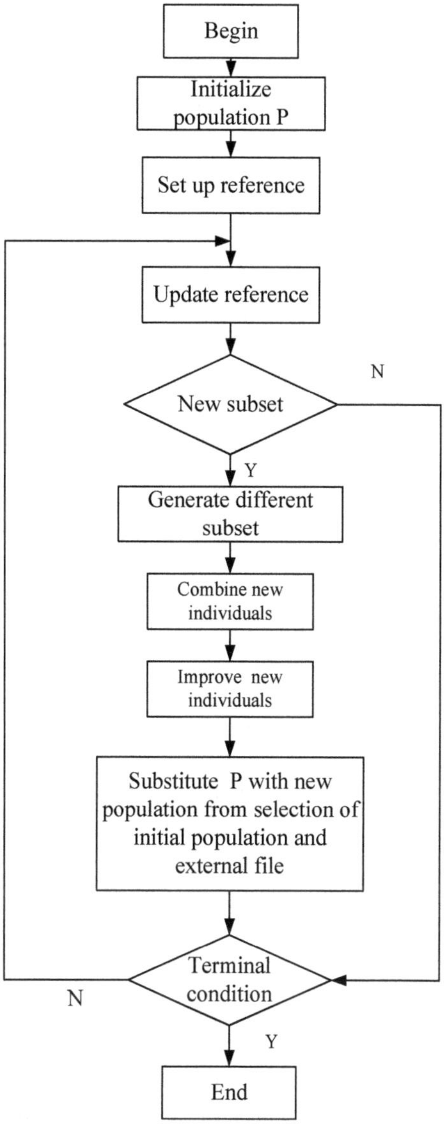

Fig. 6.36 The flowchart of AbYSS

leader is selected by a binary tournament based on the crowding value of the leader, and the maximum size of the set of leaders is equal to the swarm F. Wu et al. [19] 7181 size. The set of the leaders and the corresponding crowding values are updated after each generation. This algorithm is featured by that if the size of the set of leaders is greater than the maximum allowed, only the best leaders are retained based on their crowding value, and the others are eliminated. The detailed process of OMOPSO could be found in the study by [20].

OMOPSO process:

(1) Initialize population.
(2) Update the particle velocity according to Eq. 6.27.

$$v_i(t) = Wv_i(t-1) + C_1 r_1 (x_{pbesti} - x_i(t)) + C_2 r_2 (x_{gbest} - x_i(t)), \tag{6.27}$$

where

$v_i(t)$	the velocity of particle i after update
x_{pbesti}	Extremum of individual
x_{gbest}	Global extremum
r_1, r_2	Random number, r_1, $r_2 = \text{random}(0.0, 0.1)$
C_1, C_2	Constants control update of extremum, C_1, C_2 = random(1.5, 2.0).

(3) Update extremum of individual, if current position of particle i dominates extremum of others, then update extremum of others.
(4) Update external assemble, insert non-dominated solution into external assemble according to ε-dominated relationship.
(5) Update global extremum until condition 2 is satisfied.

6.4.1.4 SPEA2 Algorithm

SPEA (Zitzler and Thiele 1999) has been widely used in many applications. However, some shortcomings such as inaccuracy assignment of the fitness and the poor diversity of the resolutions are reported in previous research. SPEA2 Zitzler et al. [21] is proposed to overcome these shortcomings. In SPEA2, the fine-grained fitness assignment strategy is used to replace the former one and a nearest neighbor density estimation technique is incorporated to allow a more precise guidance of the search process. Additionally, an enhanced archive truncation method is adopted to guarantee the preservation of boundary solutions.

Compared with SPEA, SPEA2 is faster in convergence rate and the Pareto optimal solutions obtained by it have a more uniform distribution. So far, the integral performance of SPEA2 has improved significantly. The detailed process of SPEA2 is presented in reference Zitzler et al. [21].
Chinese version: %%多目标寻优算法的对比%%
English translation: Comparison of multi-objective optimization algorithms.

6.4.2 Introduction and Experiment Setup

Low NO_x combustion usually leads to high carbon content in fly ash and furthermore affects the boiler efficiency. It is nontrivial to find operating conditions with the simultaneous minimized NO_x emissions and maximized carbon burnout. In the multi-objective optimization of coal-fired boiler, it is expected that neither NO_x emissions nor carbon burnout will be aggravated, but at least one of NO_x emissions and carbon burnout should be improved the so-called Pareto-optimal solutions.

The experiments were carried out in a 300 MW tangentially fired dry bottom boiler with large dual furnaces of 17 m × 8.475 m cross section and 45.5 m high. The boiler is one of the units of the JianBi power plant, manufactured by Shanghai Boiler Co. Ltd. in the 1980s. Four primary air ports and six secondary air ports (represented by A–F) were arranged on each corner of the furnace, as shown in Fig. 6.37. Four medium-speed coal pulverizers supplied coal for combustion. Coal–air two-phase flow was directed at the circumference of an imaginary horizontal circle of 500 mm diameter at the furnace center. The bituminite was combusted by the concentric firing system. The characteristics of the combusted coal are listed as follows: ash content 11.06 wt%, moisture content 13.60 wt%, volatile content 31.96 wt%, and heating value 23.49 MJ/kg.

A total of 119 tests have been conducted with this boiler to investigate the effects of boiler parameters on NO_x emissions and carbon burnout characteristics. During the experiments, NO_x concentrations were continuously monitored at the boiler outlet prior to the air heater by using continuous emission monitoring system (Rosemount, Emerson Process Ltd.). The maximum NO_x emissions reported for the

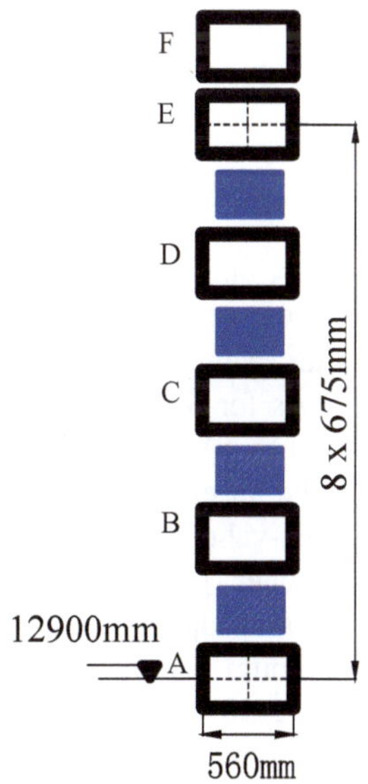

Fig. 6.37 Burner arrangement of the boiler

old-designed utility boiler was up to 404.65 ppm. The unburned carbon was measured through the solid samples retracted from the flue gas and was varied from 1.51 to 2.96%. In all tests, coal quality remained constant, and other parameters of the boiler varied in the following ranges: the primary air velocities, 24.11–29.8 m/s; secondary air velocities at the A–E elevations, 24.69–39.03 m/s and F elevation of secondary air velocities 2.19–9.87 m/s due to a restriction in the air duct arrangement; boiler load, 243.19–331.03 MW.

6.4.3 Modeling NO_x Emissions and Carbon Burnout

As a modeling technique, the Support Vector Machine (SVM) methodology was used to produce mathematical expressions using a set of input–output data (Zheng et al. [23]). Moreover, SVM could be extended to solve nonlinear regression problems by introducing the insensitive loss function ε, called support vector regression (SVR). The complete SVR equations could be obtained in Ref. Smola and Scholkopf (2004).

In this chapter, NO_x emissions and carbon burnout models were built by SVR separately as mentioned in Sect. 2.1. Identical input parameters were used for both models. The parameters were boiler load, speeds of the four mills, coal quality, four levels of primary air velocities, and six levels of secondary air velocities, respectively. The oxygen concentration in the flue gas and the total air flow rate are also important parameters to the boiler, however, they are not independent design variables and their contributions to NO_x emissions formation and carbon burnout, have been reflected by primary and secondary air velocities (Zheng et al. [23]).

Among the 119 cases obtained in the experiments, 89 cases were chosen as the training subset, the remaining as the testing subset. In order to eliminate the unit influence upon various operating parameters of the boiler, the inputs were scaled into the range [−1, 1] through Eq. (6.28).

$$x_{\text{scal}} = \frac{x_{\text{input}} - \text{range}_{\min}}{\text{range}_{\max} - \text{range}_{\min}} \times 2 - 1, \tag{6.28}$$

where $\text{range}_{\max}$ and $\text{range}_{\min}$ are upper and lower limits of the test range mentioned in next section, respectively. x_{input} and x_{scal} are operating parameters before and after scaling, respectively.

As for SVR, the parameters (g, C) have a significant influence on the models built by SVR, and should be set carefully. The grid search method has been proved to be an effective way to determine (g, C) [22]. Additionally, the application of cross-validation in searching the optimal (g, C) could avoid over fitting, and obtain a better generalization ability for SVR models. In this study, the fivefold cross-validation combined with grid search method was used to determine the optimum parameters. Pairs of exponentially growing sequences were tried, i.e., g = 2–10, 2–9.9 … 23; C = 20, 20.1 … 210. The optimal (g, C) thus obtained should reduce the MRE in the samples after fivefold cross-validation to a minimum.

In the present study, the optimal SVR parameters (g, C) for NO_x emissions model and carbon burnout model are (2.64, 34.30) and (0.47, 238.86), respectively. With the optimal parameters, the SVR models for NO_x emissions and carbon burnout could be built easily. The comparison results between the SVR models and the measurement data are present in Figs. 6.38 and 6.39. The average errors of the NO_x emission model and carbon burnout model were 1.62 and 3.79%, respectively. As for the NO_x model, the maximum error was 11.32% in the 119th case and 92% of the cases were less than 6%; for the carbon burnout model, the maximum error was 8.37% in the 27th case and 91% of them were less than 6%. It could be concluded from the results that the predicted values showed comparatively good agreement with the measured values and the SVR models were feasible and reliable.

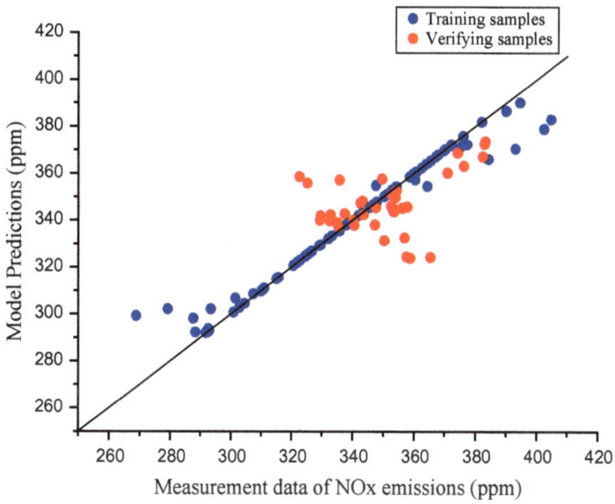

Fig. 6.38 Modeling error of NO_x emissions model built by SVR

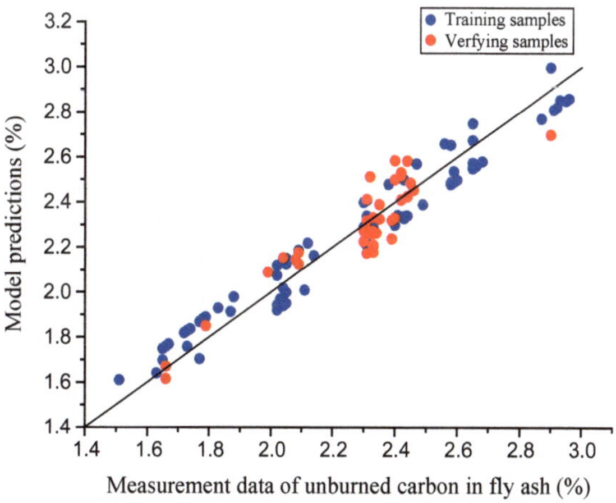

Fig. 6.39 Modeling error of carbon burnout model built by SVR

6.4.4 *Performance Metrics of Pareto Solution*

In multi-objective optimization, several performance metrics have been adopted to evaluate the algorithms. Among them, evaluation of the solution set obtained is essential, and the obtained solution set was expected as: (1) proximity to the true Pareto-optimal front, the closer to the true Pareto-optimal front of the solution set, the more accuracy is; (2) breadth against the entire true Pareto-optimal front; (3) equal distribution within the Pareto-optimal front.

In this chapter, due to the true Pareto-optimal front is unknown for the complexity of the boiler, the ratio of non-dominated individuals (RNI) was used to evaluate the accuracy of the obtained set, and cover rate was used to evaluate the breadth of solution set and how evenly the solutions distributed were evaluated by the cover rate.

6.4.4.1 The Ratio of Non-dominated Individuals (RNI)

This performance measure is derived from comparing two solutions which are derived by two methods. By RNI, the accuracy of the solutions can be compared.

The definition of RNI could be detailed in the following: supposed S_1 and S_2 were the solution sets obtained by two different methods and they could be combined to form a new solution set S_U. From S_U, the non-dominated solutions were selected to form S_P. Then, the ratios of S_1 and S_2 against S_P were calculated, respectively. The ratio closer to 100% demonstrated the corresponding method outperform the other considering the accuracy of the solutions.

6.4.4.2 Cover Rate

The diversity of the solutions should be discussed. Generally, the distribution of the solutions could be evaluated through generalized covariance value. However, covariance value was not suitable for the solutions having multiple peaks. Therefore, cover rate was used to diversity of the solutions instead of covariance.

The following procedures should be taken to derive the cover rate. First, the maximum and the minimum values of were obtained by searching one objective function f_k. Second, the distance between the minimum and the maximum was divided into N divisions, where N was any positive integer number. Third, the number of the area containing Pareto-optimum individuals was counted as N_k. Fourth, the cover ratio CR_k against the objective f_k could be obtained by Eq. (6.29).

$$\mathrm{CR}_k = \frac{N_k}{N} \tag{6.29}$$

Fifth, the above steps were conducted for every object function. Finally, the cover rate CR could be obtained by Eq. (6.30), where n was the number of the objective functions.

$$\mathrm{CR} = \frac{1}{n}\sum_{k=1}^{n} \mathrm{CR}_k. \tag{6.30}$$

The maximum cover rate is 1 and the minimum is 0. The closer the value is to 1, the more evenly distributed the solution set will be. In the present study, the Pareto solutions obtained by each algorithm will show 100 point for one object function, so the number of divisions will be 100.

6.4.5 *Parameter Settings for Various Algorithms*

A set of parameter settings were chosen to guarantee a fair comparison of the algorithms. An internal population size equal to 100 was adopted by two evolutionary algorithms (SPEA2 and MOCell). OMOPSO was configured with 100 particles and with a maximum number of 100 leaders. AbYSS used a population size of 20, which was also the size of the reference Nebro et al. [14] set, and the size of external archive was 100. In MOCell, a toroidal grid of 100 individuals (10 · 10) was chosen to structure the population and an archive of 100 individual was used. Simulated binary crossover (SBX) and polynomial mutation operators were respectively adopted as the crossover operator and mutation operator by SPEA2 and MOCell. The distribution indices for the operators were $\eta_c = 20$ and $\eta_m = 20$, and probabilities $P_c = 0.9$ and $P_m = 1/n$, for crossover and mutation respectively, where n was the number of decision variables. Two types of mutation operators, uniform and nonuniform were used by OMOPSO. For AbYSS, polynomial mutation was used in the local search procedure and SBX was used as the solution recombination method. For all the algorithms, 25,000 function evaluations were required to obtain 100 resolutions which were sufficient for convergence for the problem here.

6.4.6 *Performance Comparisons*

During the experiments, both NO_x emissions and unburned carbon in fly ash of the 53rd case were returned in high values (352.8 ppm, 2.09%) compared with other cases. Therefore, the optimization procedures were executed on this case.

In this paper, all calculations were performed on a 2.01 GHz, AMD Athlon 64 X2 Dual PC with 2 GB of RAM under Window XP (32 bit). All the above algorithms were coded in the Java language environment version 1.6.0 u6. The 100 resolutions obtained by SPEA2 costed 13469 ms CPU time, MOCell 3781 ms,

AbYSS 2391 ms, and OMOPSO only 2281 ms. It means that OMOPSO is the fastest algorithm in the experiments, and AbYSS and MOCell are slightly slower than MOCell. SPEΛ2 is the slowest and its execution time is one order higher than the others.

In order to calculate the cover rate of each algorithm, the maximum and minimum values of each object function should be assigned. The boiler load was determined by operating condition, and all the optimization procedures were performed on the 53rd case with the boiler load of 303.1 MW. The maximum and minimum values of NO_x emissions model were 375.40 and 323.02 ppm. These were equal to the maximum and the minimum values in the experiments where the boiler load was 303.1 MW. The maximum value of carbon burnout model was 2.4, which was also equal to the maximum value of unburned carbon in fly ash in the experiments with the boiler load 303.1 MW. The minimum value of unburned carbon in fly ash was set as 1, which is an empirical value.

The Pareto-optimal fronts obtained by each algorithm are shown in Figs. 6.40. Figure 6.41 shows the evaluation results of the obtained solution set using RNI and Fig. 6.42 shows the corresponding results using the cover rate.

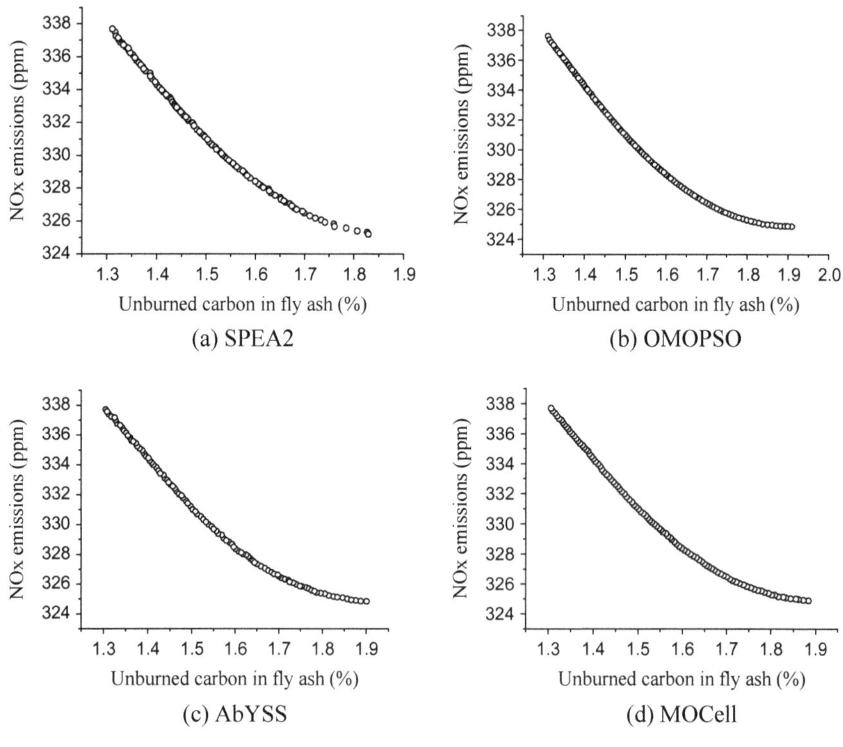

Fig. 6.40 Pareto-optimal front obtained by algorithms

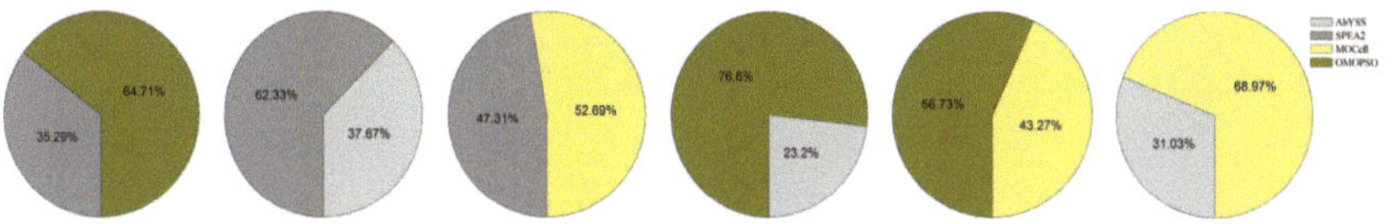

Fig. 6.41 Comparison of SPEA2, OMOPSO, AbYSS, and MOCell by RNI

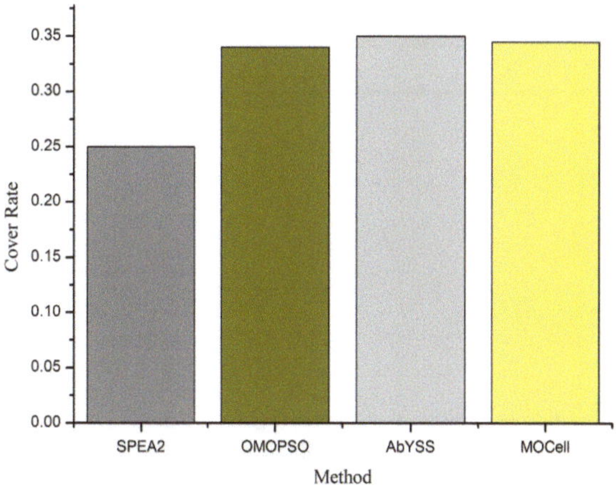

Fig. 6.42 Comparison of SPEA2, OMOPSO, AbYSS, and MOCell by cover rate

As shown in Fig. 6.40, the NO_x emissions optimized by the algorithms vary in the following range: 325.2–337.7 ppm by SPEA2, 324.9–337.6 ppm by OMOPSO, 324.8–337.7 ppm by AbYSS, 324.9–337.7 ppm by MOCell; the corresponding unburned carbon in fly ash optimized vary in the following range: 1.3–1.8%, 1.3–1.9%, 1.3–1.9% and 1.3–1.9%. Due to the large size of the result, only the resolutions with small NO_x emissions obtained by each algorithm are listed in Table 6.16. From Table 6.16, it could be concluded that the NO_x emissions and unburned carbon in fly ash after optimization have been reduced to different degrees. Moreover, the trend of air distribution obtained by each algorithm is almost the same. The primary air velocities change slightly, as they are mainly used to transport the pulverized coal, which is limited by the mill feeder and boiler load (Zheng et al. 23). Compared with primary air velocities, the secondary air velocities change significantly. The distribution of the A–E levels of secondary velocity after optimization is closer to inverse tower like. This kind of air distribution scheme makes the secondary air more rigid and enhances its ability to suspend the fuel powder, which reduces the unburned carbon in fly ash and in turn reduces the thermal losses of the boiler, a conclusion consistent with the operation experience. Additionally, this kind of air distribution scheme forms high fuel/air stoichiometric ratio at lower location of the furnace resulting in lower NO_x emissions. The analyses demonstrate the resolutions obtained by the three methods are reasonable, and all the three multi-objective methods are feasible.

Table 6.16 Comparison of the fifty-third case before and after executing optimizations

	NO_x emissions (ppm)	Unburned carbon in fly ash (%)	Load (MW)	Primary air velocities (m/s)				Secondary air velocities (m/s)					
				A	B	C	D	A	B	C	D	E	F
Field data before optimization	352.8	2.1	303.1	26.4	28.0	26.4	27.3	32.5	29.8	33.5	31.3	33.0	2.9
SPEA2	325.2	1.8	303.1	26.2	28.2	26.7	27.1	31.2	28.9	24.8	32.0	35.4	7.2
OMOPSO	324.9	1.9	303.1	26.2	28.1	26.9	27.1	31.1	28.4	24.8	31.9	34.8	7.4
AbYSS	324.8	1.9	303.1	26.2	28.1	26.9	27.1	31.3	28.6	24.9	31.8	34.9	7.4
MOCell	324.9	1.9	303.1	26.2	28.1	26.8	27.1	31.1	29.0	24.7	31.7	35.0	7.3

As shown in Fig. 6.41, the accuracy of the solution set obtained by OMOPSO is the best, which is better than the others. Additionally, the accuracy of the solution set obtained by AbYSS is the lowest. From Fig. 6.42, it could be concluded that the range covered of the interval of the solution set obtained by AbYSS is superior to others (the cover rate of SPEA2, OMOPSO, AbYSS and MOCell are 30, 34, 34.5 and 35%, respectively). In addition, the range covered and closeness of the interval of the solution set obtained by SPEA2 is the lowest.

In summary, it is concluded that the OMOPSO algorithm, for the cases studied in this chapter, shows much better results than other optimization approaches in terms of accuracy. At the same time, it takes less CPU time to execute than others. Considering the diversity, AbYSS is the best, but is just slightly better than OMOPSO and MOCell.

6.4.7 Conclusion

This study aims at comparing various optimization algorithms applied in the coal-fired utility boiler for multi-objective optimization. SVR was employed to build the models for NO_x emissions and carbon burnout, and the predicted NO_x emissions and carbon burnout from the SVR models showed good agreement with the measurement. Three multi-objective optimization algorithms were, respectively, presented for realizing Pareto-based multi-objective optimization of the coal-fired boiler by carefully setting the operational parameters. The results show that the method combining SVR and multi-objective algorithms could effectively reduce NO_x emissions and unburned carbon in fly ash of the coal-fired boiler simultaneously. Detailed comparisons among the three algorithms were made in terms of accuracy, diversity, and the convergence rate. The results show that OMOPSO algorithm is the best in terms of the accuracy and convergence rate. AbYSS has a better diversity but the worst accuracy. SPEA2 is the slowest algorithm and its execution time is one order higher than other algorithms. Therefore, OMOPSO and MOCell are the proposed algorithms for online multi-object optimization of coal-fired boilers.

6.5 Summary

In this section, optimization algorithms combining with SVM and ANN is introduced. GA, ACO, and PSO are commonly used optimization algorithm for combustion optimization. They are adopted in model and operation parameters optimization to gain the minimum goal for a different function.

Different optimization samples based on SVM and ANN combining optimization are detailed discussed. Also, multi-objective optimization is introduced, muti-objective algorithm is suitable for the industrial boiler. It is concluded that the OMOPSO algorithm, for the cases studied in this chapter, shows much better results than other optimization approaches in terms of accuracy.

References

1. Kamas J, Keeler J. Predictive emissions monitoring systems: a low-cost alternative for emissions monitoring (in cement industry). In: Proceedings of the 37th IEEE cement industry technical conference, San Juan, Puerto Rico; 1995. p. 497–509.
2. Hua Z, Wang Y, Xu X, Zhang B, Liang L. Predicting corporate financial distress based on integration of support vector machine and logistic regression. Expert Syst Appl. 2007;33 (2):434–40.
3. Specht DF. A general regression neural network. IEEE Trans Neural Netw. 1991;2(6): 568–70.
4. Chen W-H. A simplified model of predicting coal reaction in a partial oxidation environment. Int Commun Heat Mass Transf. 2007;34:623–9.
5. Huang C-L, Wang C-J. A GA-based feature selection and parameters optimization for support vector machines. Expert Syst Appl. 2006;31(2):231–40.
6. Hsu CW, Chang CC, Lin CJ. A practical guide to support vector classification; 2003.
7. Hsu CW, Lin CJ. A simple decomposition method for support vector machine. Mach Learn. 2002;46(1–3):219–314.
8. Sivagaminathan RK, Ramakrishnan S. A hybrid approach for feature subset selection using neural networks and ant colony optimization. Expert Syst Appl. 2007;33(1):49–60.
9. Radl BJ. Neural networks prove effective at NO_x reduction. Modern Power Syst. 2000;20 (5):59–61.
10. Socha K. ACO for continuous and mixed-variable optimization. In: Dorigo M et al, editors. ANTS 2004, LNCS, vol. 3172. Berlin/Heidelberg: Springer-Verlag; 2004. p. 25–36.
11. Socha K, Blum C. An ant colony optimization algorithm for continuous optimization: application to feed-forward neural network training. Neural Comput Appl. 2007;16(3): 235–47.
12. Zhou H, Cen KF, Fan JR. Multi-objective optimization of the coal combustion performance with artificial neural networks and genetic algorithms. Int J Energy Res. 2005;29(6):499–510.
13. Elbeltagi E, Hegazy T, Grierson D. Comparison among five evolutionary-based optimization algorithms. Adv Eng Inform. 2005;19(1):43–53.
14. Nebro AJ, Durillo JJ, Luna F, Dorronsoro B, Alba E. A cellular genetic algorithm for multiobjective optimization. In: Proceedings of NICSO, Granada, Spain; 2006.
15. Glover F, Laguna M, Marti R. Fundamentals of scatter search and path relinking. Control Cybern. 2000;29(3):653–84.
16. Glover F. A template for scatter search and path relinking. Lecture notes in computer science. Springer Verlag; 1997.
17. Kennedy J, Eberhart RC. Particle swarm optimization. In: Proceedings of the 1995 IEEE international conference on neural networks, Piscataway, New Jersey; 1995. p. 1942–8.
18. Deb K, Partap A, Agarwal S, Meyarivan T. A fast and elitist multi-objective genetic algorithm: NSGA-II. IEEE Trans Evol Comput. 2002;6(22):182–97.
19. Wu F, Zhou H, Zhao J, Cen K. A comparative study of the multi-objective optimization algorithms for coal-fired boilers[J]. Expert Systems with Applications. 2011;38(6): 7179–7185.
20. Sierra MR, Coello CAC. Improving PSO-based multi-objective optimization using crowding, mutation and epsiv-dominance. In: Third international conference on evolutionary multi-criterion optimization (EMO 2005), Guanajuato, Mexico; 2005. p. 505–19.

21. Zitzler E, Deb K, Thiele L. Comparison of multiobjective evolutionary algorithms: empirical results. Evol Comput. 2000;8(2):173–95.
22. Zheng LG, Zhou H, Cen KF, Wang CL. A comparative study of optimization algorithms for low NO_x combustion modification at a coal-field utility boiler. Expert Syst Appl. 2009;36 (2):2780–93.
23. Zheng L, Zhou H, Wang C, et al. Combining support vector regression and ant colony optimization to reduce NO_x emissions in coal-fired utility boilers. Energy Fuels. 2008;22 (2):1034–40.

Chapter 7
Online Combustion Optimization System

Abstract In this chapter, the online combustion optimization system is exposed. First, the demand of the software system for the practical combustion optimization and the need of the local optimization are summarized, such as data detection requirements, quickness and accuracy requirements, requirements of different optimization goal, requirements online self-learning, parameter optimization limit requirements, fault tolerance requirements, alarm requirements, compatibility of off-line data processing and optimizing and so on. Second the instruments or sensors for online combustion optimization system are introduced. Then, the online SVM algorithm is presented, mainly about derivation of the Incremental Relations, AOSVR Bookkeeping Procedure, efficiently updating the R Matrix, initialization of the Incremental Algorithm and Decremental Algorithm. In addition, there are three main functions with different modules of online combustion optimization system. They are, respectively, online monitoring and alarm function, online optimization and self-learning function, off-line modeling, and optimization function. Finally the application of online combustion optimization system is discussed.

Keywords Online combustion optimization system · Data detection requirements
Quickness and accuracy requirements · Requirements of different optimization goal
Requirements online self-learning · Parameter optimization limit requirements
Fault tolerance requirements · Alarm requirements · Compatibility of off-line data processing and optimizing · Instruments or sensors · Online SVM algorithm
Derivation of the Incremental Relations · AOSVR Bookkeeping Procedure
Efficiently updating the R Matrix · Initialization of the Incremental Algorithm
Decremental Algorithm · Online monitoring and alarm function
Online optimization and self-learning function · Off-line modeling and optimization function · The application of online combustion optimization system

H. Zhou and K. Cen, *Combustion Optimization Based on Computational Intelligence*, Advanced Topics in Science and Technology in China,
https://doi.org/10.1007/978-981-10-7875-0_7

7.1 Introduction

Combustion in boiler is a complex process, which is related to combustion condition and design factors, and design factors cannot be changed for a boiler in operation. So the optimal combustion can only be implemented via combustion operation adjustment, and the adjustable parameters that can affect the combustion state include: primary air, secondary air, oxygen, coal, and coal fineness, and those parameters also should be suitable for boiler load change and coordinated. In the actual boiler operation, the personnel refer mostly to his own experience, it is almost impossible to make boiler combustion in the best status. Guiding the practical operation by using the method of the combustion adjustment test can improve the quality of the operation of the boiler, but burning adjustment test is time consuming and the working condition of the boiler is limited, so combustion optimization is needed.

Through our actual inspection during boiler combustion adjustment and communication with power plant operators, we summarize the demand of the software system for the practical combustion optimization and the need for the local optimization as follows.

7.1.1 Data Detection Requirements

In boiler combustion operation, the operator cannot get important evaluation indicator of combustion in time. It is necessary to transmit the related combustion information to the operator, and this function is included in the combustion optimization software. Although the instrument can be used to deliver the information as wind speed, oxygen, but the parameters reflecting boiler efficiency and NO_x emission cannot be monitored immediately. As a result, the operator cannot achieve an effective evaluation of combustion state, which is needed to be improved.

7.1.2 Quickness and Accuracy Requirements

The wind, with the powder and oxygen content in boiler combustion should be changed according to changes of the load and coal, so to maintain a good combustion state real-time guidance is required, which means to provide the targeted specific adjustment according to the combustion. The plan should be provided timely, otherwise, a given scheme has lost its meaning when conditions change before the adjustment. On the other hand, the adjustment should be accurate, which refers to evaluation after the adjustment of the adjustment scheme given for combustion state is accurate, we can achieve the desired goal according to the adjustment. This is mainly associated with the prediction ability of the model.

7.1.3 Requirements for Different Optimization Goals

Different adjustments for sundry combustion targets according to the actual condition are implemented, and different targets are interactional (like boiler efficiency and NO_x emission), sometimes requirements of all the optimization cannot be satisfied at the same time. When each evaluation meets the requirements, the specific situation should be focused. This optimization strategy requires that the operation parameters should be linked with different combustion evaluations to establish independent models, and each model can be combined organically, optimization goal is set according to the different conditions and requirements. The flexibility of optimal goal setting for boiler combustion optimization is in demand.

7.1.4 Requirements Online Self-Learning

Boiler combustion changes complexly due to the variation of load and coal type. With the growth of the running time, the changes of a characteristic of the combustion equipment will also make changes in the burning state, and operation stoppage will directly lead to changes in combustion. Some faults in the operation cannot be ruled out temporarily, this may affect combustion in the long term. At this moment fault factors should be considered in combustion optimization, the new situation, the conditions, and equipment information need to be learned, otherwise optimization will not be able to adapt to real conditions.

7.1.5 Parameter Optimization Limit Requirements

Combustion optimization organizes boiler operation according to load, coal type, and the air, oxygen corresponding, leading to high boiler efficiency and low pollution. The wind, powder, and oxygen parameters are limited to a certain range according to the design or operation experience, otherwise larger optimization adjustment range will increase the time of optimization. On the other hand, the model is calculated based on pure mathematics, the best mathematical optimal value could not be realized without imitation.

7.1.6 Fault Tolerance Requirements

As a software, combustion optimization system should have fault tolerance. The need of fault tolerance is mainly for unexpected events or failures. These events may result in anomalies or errors of some data. If there is no fault tolerance, optimization software cannot run regularly and even the system crashes. Fault tolerance makes the combustion optimization possible in incomplete data and bad data situation. It is a great supplement to the integrity of the combustion optimization of the boiler.

7.1.7 *Alarm Requirements*

Alarm ensures the safety of boiler combustion. It is difficult for workers to monitor a lot of potential failures in the boiler combustion at the same time. This requires the real-time online monitoring system to achieve the alarm. To ensure the safety of the boiler combustion especially when the condition changes. Online combustion optimization of the boiler should not endanger the safety of combustion. And, it can detect multiple parameters in real time. It can alarm on time when an abnormality is discovered. Let workers deal with the failures to ensure the safety operation. This is the need for safe operation.

7.1.8 *Compatibility of Off-line Data Processing and Optimizing*

The new units need to take commissioning tests before being put into operation. After long-term operation, the boiler needs overhaul. It needs combustion debugging before running again. The system cannot obtain the online data because boiler combustion characteristics are often unknown or changed when compared with the previous, or due to some objective conditions. So, it needs off-line modeling and optimizing of experimental data. On one hand, it can provide data which can support commissioning tests. On the other hand, data and models obtained by experiments for the next step of online self-learning foundation for modeling and optimization make the foundation of online self-learning of modeling and optimization. Compatible off-line test data processing and optimization provide a convenience for the operational and technical workers and online optimization.

7.2 Instruments or Sensors for Online Combustion Optimization System

The flame monitor (Fig. 7.1) is developed for coal identification (in Chap. 5). It can achieve the measurement of flame temperature and other characteristics. The flame monitor consists of a water cooling tube, a camera lens, a beam splitter, a signal-processing board with three photodetectors (for infrared (IR), visible (VI), and ultraviolet (UV) signals, separately), a digital camera and an embedded motherboard.

The camera lens, protected by cooling water and central air flow, was used to guide the light of flame from the furnace to the beam splitter. The beam splitter divides the light into two beams. One beam is captured by the digital camera to obtain flame images. The other beam is transmitted to the photodetectors on the signal-processing board to obtain light intensity signals. The flame images and light intensity signals

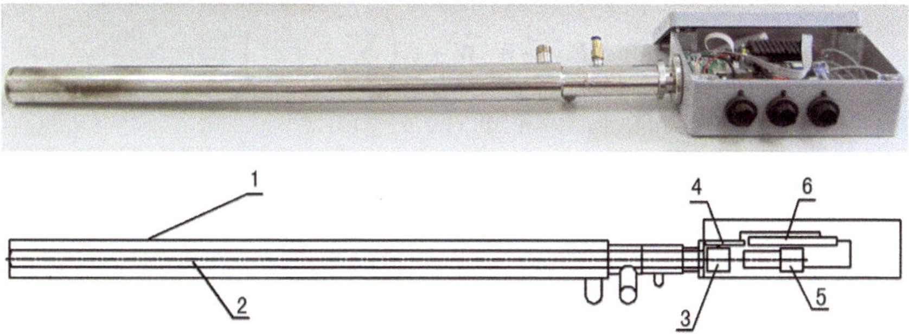

Fig. 7.1 Flame monitoring system: 1: Water cooling tube, 2: Camera lens, 3: Beam splitter, 4: Photodetectors and signal-processing board, 5: Digital camera, 6: Embedded motherboard

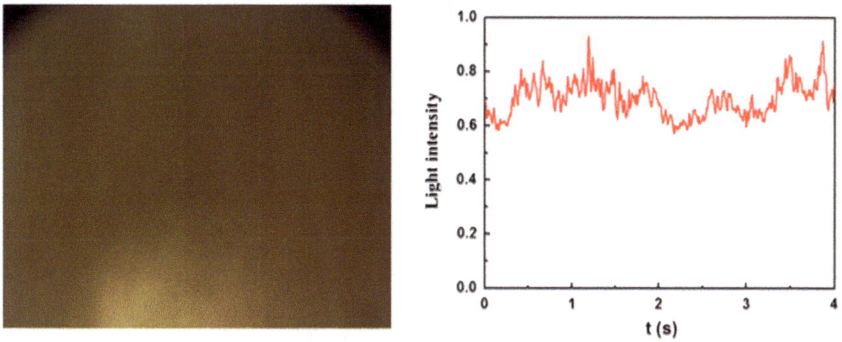

Fig. 7.2 An example of flame image and light intensity signal

(an example is shown in Fig. 7.2) were received by the embedded motherboard. Application software on the embedded motherboard processes the flame images and light intensity signals to extract spatial and temporal features of the flame.

7.3 Online SVM Algorithm

7.3.1 Algorithm Introduction

Given a training set $T = \{(x_i, y_i), i = 1, 2, \ldots, l\}$, where $x_i \in R^N$ and $y_i \in R$, we construct a linear regression function,

$$f(x) = W^T \Phi(x) + b \tag{7.1}$$

on a feature space F. Here, W is a vector in F, and $\Phi(x)$ maps the input x to a vector in F. The W and b in Eq. 7.1 are obtained by solving an optimization problem as follows:

$$\begin{aligned}
\min \quad & P = \frac{1}{2} W^T W + C\left(\sum_{i=1}^{l} \xi_i + \xi_i^*\right) \\
\text{s.t.} \quad & y_i - (W^T \Phi(x) + b) \le \varepsilon + \xi_i \\
& (W^T \Phi(x) + b) - y_i \le \varepsilon + \xi_i^* \\
& \xi_i, \xi_i^* \ge 0, \; i = 1, 2, \ldots, l
\end{aligned} \tag{7.2}$$

The optimization criterion penalizes data points whose y-values differ from $f(x)$ by more than ε. The slack variables, ξ_i and ξ_i^*, correspond to the size of this excess deviation for positive and negative deviations, respectively, as shown in Fig. 7.1.

Introducing Lagrange multipliers α, α^* and η, η^*, we can write the corresponding Lagrangian as (Fig. 7.3)

$$\begin{aligned}
L_P = & \frac{1}{2} W^T W + C\left(\sum_{i=1}^{l} \xi_i + \xi_i^*\right) - \sum_{i=1}^{l} \eta_i \xi_i + \eta_i^* \xi_i^* \\
& - \sum_{i=1}^{l} \alpha_i (\varepsilon + \xi_i + y_i - W^T \Phi(x) - b) \\
& - \sum_{i=1}^{l} \alpha_i^* (\varepsilon + \xi_i^* + W^T \Phi(x) + b - y_i) \\
\text{s.t.} \quad & \alpha_i, \alpha_i^*, \eta_i, \eta_i^* \ge 0, \; i = 1, 2, \ldots, l
\end{aligned}$$

This, in turn, leads to the dual optimization problem:

$$\begin{aligned}
\min \quad & D = \frac{1}{2} \sum_{i=1}^{l} \sum_{j=1}^{l} Q_{ij} (\alpha_i - \alpha_i^*)(\alpha_j - \alpha_j^*) + \varepsilon \sum_{i=1}^{l} (\alpha_i + \alpha_i^*) - \sum_{i=1}^{l} y_i (\alpha_i - \alpha_i^*) \\
\text{s.t.} \quad & 0 \le \alpha_i, \alpha_i^* \le C \quad i = 1, 2, \ldots, l \\
& \sum_{i=1}^{l} (\alpha_i - \alpha_i^*) = 0
\end{aligned} \tag{7.3}$$

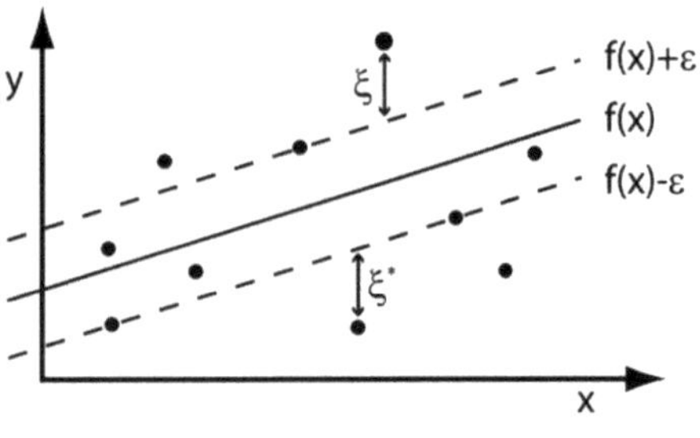

Fig. 7.3 The ε-insensitive loss function and the role of slack variables ξ and ξ^*

where $Q_{ij} = \Phi(x_i)^T\Phi(x_j) = K(x_i) \bullet (x_j)$, here $K(x_i) \bullet (x_j)$ is a kernel function. Given the solution of Eq. 7.3, the regression function 7.1 can be written as

$$f(x) = \sum_{i=1}^{l}(\alpha_i - \alpha_i^*)K(x_i, x) + b \tag{7.4}$$

The Lagrange formulation of Eq. 7.3 can be represented as

$$\begin{aligned} L_D = &\frac{1}{2}\sum_{i=1}^{l}(\alpha_i - \alpha_i^*)(\alpha_j - \alpha_j^*) + \varepsilon\sum_{i=1}^{l}(\alpha_i + \alpha_i^*) - \sum_{i=1}^{l}y_i(\alpha_i - \alpha_i^*) \\ &- \sum_{i=1}^{l}(\delta_i\alpha_i + \delta_i^*\alpha_i^*) + \sum_{i=1}^{l}[u_i(\alpha_i - C) + u_i^*(\alpha_i^* - C)] + \zeta\sum_{i=1}^{l}(\alpha_i + \alpha_i^*) \end{aligned}, \tag{7.5}$$

where $\delta_i^{(*)}$, $u_i^{(*)}$ and ζ are the Lagrange multipliers. Optimizing this Lagrangian leads to the Karush–Kuhn–Tucker (KKT) conditions:

$$\begin{aligned} &\frac{\partial L_D}{\partial \alpha_i} = \sum_{j=1}^{l} Q_{ij}(\alpha_j - \alpha_j^*) + \varepsilon - y_i + \zeta - \delta_i^* + u_i = 0 \\ &\frac{\partial L_D}{\partial \alpha_i^*} = \sum_{j=1}^{l} Q_{ij}(\alpha_j - \alpha_j^*) + \varepsilon - y_i - \zeta - \delta_i^* + u_i^* = 0 \\ &\delta_i^{(*)} \geq 0, \quad \delta_i^{(*)}\alpha_i^* = 0 \\ &u_i^{(*)} \geq 0, \quad u_i^{(*)}(\alpha_i^* - C) = 0 \end{aligned} \tag{7.6}$$

Note that ζ in Eq. 7.6 is equal to b in Eqs. 7.1 and 7.4 at optimality.

According to the KKT conditions in Eq. 7.6, at most one α_i^* and α_i of will be nonzero, and both are nonnegative. Therefore, we can define a coefficient difference θ_i as

$$\theta_i = \alpha_i - \alpha_i^* \tag{7.7}$$

and note that θ_i determines both α_i^* and α_i.

Define a margin function $h(x_i)$ for the i th sample x_i as

$$h(x_i) \equiv f(x_i) - y_i = \sum_{j=1}^{l} Q_{ij}\theta_j - y_j + b \tag{7.8}$$

Combining Eqs. 7.6, 7.7, and 7.8, we can obtain

$$\begin{cases} h(x_i) \geq \varepsilon & \theta_i - -C \\ h(x_i) = \varepsilon & -C < \theta_i < C \\ -\varepsilon \leq h(x_i) \leq \varepsilon & \theta_i = 0 \\ h(x_i) = \varepsilon & 0 < \theta_i < C \\ h(x_i) \leq -\varepsilon & \theta_i = C \end{cases} \tag{7.9}$$

There are five conditions in Eq. 7.9, compared to the three conditions in support vector classification (see Eq. 2 in Cauwenberghs and Poggio [1]), but like the conditions in support vector classification, they can be identified with three subsets into which the samples in training set T can be classified. The difference is that two of the subsets (E and S) are themselves composed of two disconnected components, depending on the sign of the error $f(x_i) - y_i$.

$$\text{The } E \text{ set: Error support vector}: E = \{i | \, |\theta_i| = C\} \tag{7.10}$$

$$\text{The } S \text{ set: Margin support vector: } S = \{i | \, 0 < |\theta_i| < C\} \tag{7.11}$$

$$\text{The } R \text{ set: Remaining sample: } R = \{i | \theta_i = 0\} \tag{7.12}$$

The incremental algorithm updates the trained SVR function, whenever a new sample x_c is added to the training set T. The basic idea is to change the coefficient θ_c corresponding to the new sample x_c in a finite number of discrete steps until it meets the KKT conditions, while ensuring that the existing samples in T continue to satisfy the KKT conditions at each step. In this section, we first derive the relation between the change of θ_c or $\Delta\theta_c$ and the change of other coefficients under the KKT conditions, and then propose a method to determine the largest allowed $\Delta\theta_c$ for each step. A pseudocode description of this algorithm is provided in the appendix.

7.3.2 Derivation of the Incremental Relations

Let x_c be a new training sample that is added to T. We initially set $\theta_c = 0$ and then gradually change (increase or decrease) the value of θ_c under the KKT conditions, Eq. 7.9.

According to Eqs. 7.6, 7.7, and 7.9, the incremental relation between $\Delta h(x_i)$ $\Delta\theta c$ and Δb is given by

$$\Delta h(x_i) \equiv Q_{ij}\Delta\theta_c + \sum_{j=1}^{l} Q_{ij}\Delta\theta_j + \Delta b. \tag{7.13}$$

From the equality condition in Eq. 2.3, we have

$$\theta_c + \sum_{i=1}^{l} \theta_i = 0 \tag{7.14}$$

Combining Eqs. 7.9 through 7.12, 7.12, kernel function, and 7.14, we obtain the following:

$$\begin{aligned} \sum_{j \in S} Q_{ij} \Delta\theta_j + \Delta b &= -Q_{ij} \Delta\theta_c \quad \text{where} \quad i \in S \\ \sum_{j \in S} \Delta\theta_j &= -\Delta\theta_c \end{aligned} \tag{7.15}$$

If we define the index of the samples in the S set as follows:

$$S = \{s_1, s_2, \ldots, s_{ls}\} \tag{7.16}$$

Equation 7.15 can be represented in matrix form as

$$\begin{bmatrix} 0 & 1 & \cdots & 1 \\ 1 & Q_{s_1 s_1} & \cdots & Q_{s_1 s_{ls}} \\ \vdots & \vdots & \ddots & \vdots \\ 1 & Q_{s_{ls} s_1} & \cdots & Q_{s_{ls} s_{ls}} \end{bmatrix} \begin{bmatrix} \Delta b \\ \Delta\theta_{s_1} \\ \vdots \\ \Delta\theta_{s_{ls}} \end{bmatrix} = - \begin{bmatrix} 1 \\ Q_{s_1 c} \\ \vdots \\ Q_{s_{ls} c} \end{bmatrix} \Delta\theta_c \tag{7.17}$$

That is,

$$\begin{bmatrix} \Delta b \\ \Delta\theta_{s_1} \\ \vdots \\ \Delta\theta_{s_{ls}} \end{bmatrix} = \beta \Delta\theta_c, \tag{7.18}$$

where

$$\beta = \begin{bmatrix} 1 \\ \beta_{s_1} \\ \vdots \\ \beta_{s_{ls}} \end{bmatrix} = -R \begin{bmatrix} 1 \\ Q_{s_1 c} \\ \vdots \\ Q_{s_{ls} c} \end{bmatrix}$$

where

$$R = \begin{bmatrix} 0 & 1 & \cdots & 1 \\ 1 & Q_{s_1 s_1} & \cdots & Q_{s_1 s_{ls}} \\ \vdots & \vdots & \ddots & \vdots \\ 1 & Q_{s_{ls} s_1} & \cdots & Q_{s_{ls} s_{ls}} \end{bmatrix}^{-1} \tag{7.19}$$

Define a non-S, or N, set as $N = E \cup R = \{n_1, n_2, \ldots, n_{ln},\}$. Combining Eqs. 7.9 through 7.12, 7.13, and 7.18, we obtain

$$\begin{bmatrix} \Delta h(x_{n_1}) \\ \Delta h(x_{n_2}) \\ \vdots \\ \Delta h(x_{n_{ln}}) \end{bmatrix} = \gamma \Delta\theta_c \tag{7.20}$$

where

$$\gamma = \begin{bmatrix} Q_{n_1 c} \\ Q_{n_2 c} \\ \vdots \\ Q_{n_{ln} c} \end{bmatrix} \begin{bmatrix} 1 & Q_{n_1 s_1} & \cdots & Q_{n_1 s_{ls}} \\ 1 & Q_{n_2 s_1} & \cdots & Q_{n_2 s_{ls}} \\ \vdots & \vdots & \ddots & \vdots \\ 1 & Q_{n_{ls} s_1} & \cdots & Q_{n_{ls} s_{ls}} \end{bmatrix} \beta \tag{7.21}$$

In the special case when S set is empty, according to Eqs. 7.13 and 7.14, Eq. 7.21 simplifies to $\Delta h(x_n) = \Delta b$, for all $n \in E \cup R$.

Given $\Delta\theta_c$, we can update θ_i, $i \in S$ and b according to Eq. 7.18 and update $h(x_i)$, $i \in N$, according to Eq. 7.8. Moreover, Eq. 7.9 suggests that θ_i, $i \in N$, and $h(x_i)$, $i \in S$ are constants if the S set stays unchanged. Therefore, the results presented in this section enable us to update all the θ_i and $h(x_i)$ given $\Delta\theta_c$. In the next section, we address the question of how to find an appropriate $\Delta\theta_c$.

7.3.3 AOSVR Bookkeeping Procedure

Equations 7.18 and 7.10 hold only when the samples in the S set do not change membership. Therefore, $\Delta\theta_c$ is chosen to be the largest value that either can maintain the S set unchanged or lead to the termination of the incremental algorithm.

The first step is to determine whether the change $\Delta\theta_c$ should be positive or negative. According to Eq. 7.9,

$$\mathrm{sign}(\Delta\theta_c) = \mathrm{sign}(y_c - f(x_c)) = \mathrm{sign}(-h(x_c)). \tag{7.22}$$

The next step is to determine a bound on $\Delta\theta_c$ imposed by each sample in the training set. To simplify exposition, we consider only the case $\Delta\theta_c > 0$, and remark that the case $\Delta\theta_c < 0$ is similar.

For the new sample $\boldsymbol{x}_c$, there are two cases:

Case 1: $h(x_c)$ change from $h(\boldsymbol{x}_c) < -\varepsilon$ to $h(\boldsymbol{x}_c) = -\varepsilon$, and the new sample $\boldsymbol{x}_c$ is added to the S set, and the algorithm terminates.
Case 2: If θ_c increases from $\theta_c < C$ to $\theta_c = C$, new sample $\boldsymbol{x}_c$ is added to the E set, and the algorithm terminates.
For each sample $\boldsymbol{x}_i$ in the set S,
Case 3: If θ_i change from $0<|\theta_i|<C$ to $|\theta_i| = C$, sample $\boldsymbol{x}_i$ changes from the S set to the E set. If θ_i change to $\theta_i = 0$, sample $\boldsymbol{x}_i$ changes from the S set to the R set.
For each sample $\boldsymbol{x}_i$ in the set E,
Case 4: If $h(\boldsymbol{x}_i)$ changes from $|h(\boldsymbol{x}_c)| > \varepsilon$ to $|h(\boldsymbol{x}_c)| = \varepsilon$, $\boldsymbol{x}_i$ is moved from the E set to the S set.
For each sample $\boldsymbol{x}_i$ in the set R,
Case 5: If $h(\boldsymbol{x}_i)$ changes from $|h(\boldsymbol{x}_c)|<\varepsilon$ to $|h(\boldsymbol{x}_c)| = \varepsilon$, $\boldsymbol{x}_i$ is moved from the R set to the S set.

The bookkeeping procedure is to trace each sample in the training set T against these five cases and determine the allowed $\Delta\theta_c$ for each sample according to Eq. 7.18 or 7.20. The final $\Delta\theta_c$ is defined as the one with a minimum absolute value among all the possible $\Delta\theta_c$.

7.3.4 *Efficiently Updating the* R *Matrix*

The matrix **R** that is used in Eq. 7.19,

$$R = \begin{bmatrix} 0 & 1 & \cdots & 1 \\ 1 & Q_{s_1 s_1} & \cdots & Q_{s_1 s_{ls}} \\ \vdots & \vdots & \ddots & \vdots \\ 1 & Q_{s_{ls} s_1} & \cdots & Q_{s_{ls} s_{ls}} \end{bmatrix}^{-1} \tag{7.23}$$

must be updated whenever the S set is changed. Following Cauwenberghs and Poggio [1], we can efficiently update **R** without explicitly computing the matrix inverse. When the kth sample x_{sk} in the S set is removed from the S set, the new **R** can be obtained as follows:

$$R^{\text{new}} = R_{I,I} - \frac{R_{I,k} R_{k,I}}{R_{k,k}} \tag{7.24}$$

where $I = [1\ldots kk+2\ldots S_{ls}+1]$.

When a new sample is added to S set, the new $\mathbf{R}$ can be updated as follows:

$$R^{\text{new}} = \begin{bmatrix} & & & 0 \\ & R & & \vdots \\ & & & 0 \\ 0 & \cdots & 0 & 0 \end{bmatrix} + \frac{1}{\gamma_i}\begin{bmatrix} \beta \\ 1 \end{bmatrix}\begin{bmatrix} \beta^T & 1 \end{bmatrix} \tag{7.25}$$

where β is defined as

$$\beta = -R\begin{bmatrix} 1 \\ Q_{s_1 i} \\ \vdots \\ Q_{s_{1s} i} \end{bmatrix}$$

And, γ_i is defined as

$$\gamma_i = Q_{ii} + \begin{bmatrix} 1 \\ Q_{s_1 i} \\ \vdots \\ Q_{s_{1s} i} \end{bmatrix}\beta$$

when the sample $\boldsymbol{x}_i$ was moved from E set to R set. In contrast, when the sample $\boldsymbol{x}_c$ is the sample added to S set, β can be obtained according to Eq. 7.19, and γ_i is the last element of γ defined in Eq. 7.21.

7.3.5 *Initialization of the Incremental Algorithm*

An initial SVR solution can be obtained from a batch SVR solution, and in most cases that is the most efficient approach. But it is sometimes convenient to use AOSVR to produce a full solution from scratch. An efficient starting point is the two-sample solution. Given a training set $T = \{(x_1, y_1), (x_2, y_2)\}$, with $y_1 \geq y_2$, the solution of Eq. 7.3 is

$$\begin{aligned} \theta_1 &= \max\left(0, \min\left(C, \frac{y_1 - y_2 - 2\varepsilon}{2(K_{11} - K_{12})}\right)\right) \\ \theta_2 &= -\theta_1 \\ b &= (y_1 - y_2)/2 \end{aligned} \tag{7.26}$$

The sets E, S, and R are initialized from these two points based on Eqs. 7.10 through 7.12. If the set S is nonempty, the matrix R can be initialized from Eq. 7.23. As long as S is empty, the matrix R will not be used.

7.3.6 Decremental Algorithm

The decremental (or "unlearning") algorithm is employed when an existing sample is removed from the training set. If a sample $\boldsymbol{x}_c$ is in the R set, then it does not contribute to the SVR solution, and removing it from the training set is trivial; no adjustments are needed. If, on the other hand, $\boldsymbol{x}_c$ has a nonzero coefficient, then the idea is to gradually reduce the value of the coefficient to zero, while ensuring all the other samples in the training set continue to satisfy the KKT conditions.

The decremental algorithm follows the incremental algorithm with a few small adjustments as follows:

- The direction of the change of θ_c is

$$\operatorname{sign}(\Delta\theta_c) = \operatorname{sign}(f(x_c) - y_c) = \operatorname{sign}(h(x_c)) \tag{7.27}$$

- There is no case 1 because the removed $\boldsymbol{x}_c$ needs not satisfy KKT conditions.
- The condition in case 2 becomes: θ_c changing from $|\theta_c| > 0$ to $\theta_c = 0$.

7.4 Online Combustion Optimization System

The online power plant boiler combustion optimization software is developed mainly to satisfy the eight requirements mentioned above. On account of the actual situation, we realize the following three main functions with different modules.

7.4.1 Online Monitoring and Alarm Function

The online combustion optimization system can monitor the boiler oxygen content of rear section, carbon monoxide concentrations near the waterwall in furnace and NO_x emission in real time with online gas analysis system, as well as parameters like the wind speed, the powder, oxygen in furnace from MIS system. Once abnormal data or the parameters may cause failure is found, the system will send an alarm signal, which is aimed at air/power combustion system failure and high-temperature corrosion. The system structure of online gas analysis system monitoring the boiler oxygen content of rear section, carbon monoxide concentrations near the waterwall in furnace and NO_x emission in real time is shown in Fig. 7.4.

The subsystems of online gas analysis system include gas sampling tube, heat transfer tube of sample gas, sample back system, switching device, sampling pump,

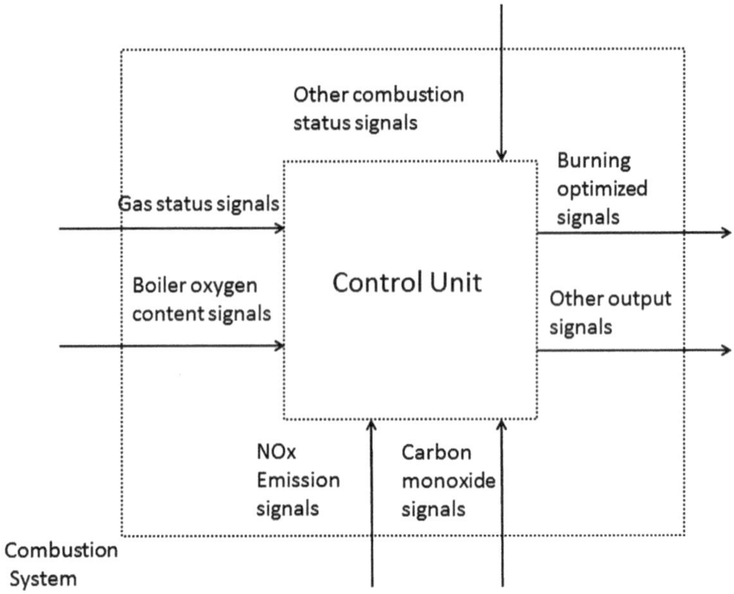

Fig. 7.4 Online gas monitoring and combustion optimization system

compressor cooler, and flue gas composition analysis instruments. The gas is exported from the furnace or air preheater, delivered to switch device after heated in the filter, then dehydrated and cooled in shrinkage machine, finally transferred into the analysis instrument, and analysis data is saved in the combustion optimization system.

7.4.2 Online Optimization and Self-Learning Function

The optimization based on genetic algorithm and boiler combustion model is implemented online according to the load and coal type, achieving better economy, environmental protection, and safety by adjustment of parameters cooperation. During combustion optimization, gas composition of the furnace and the tail regent is picked from online gas monitoring subsystems. The flowchart of online gas monitoring system and combustion optimization system is shown in Fig. 7.5. The online optimization process is shown in Fig. 7.6.

First, combustion information is collected from the online gas monitoring system and MIS, then SVM model is applied based on this information and GA is implemented to optimize the parameters to achieve the best boiler combustion.

7.4.3 Off-line Modeling and Optimization Function

Off-line modeling and optimization functions are established for power plants during the repair or after combustion adjustment test, as the combustion characteristic differs for these two conditions. So the online model built before is not suitable for optimization at present, the new model should be built according to experimental data and then optimization is implemented. Online data acquisition may not work because of all kinds of objective conditions, so the function of test data acquisition manually is provided, which is applied to building model and modification if needed.

The above three main functions and some other auxiliary functions are realized by three modules in the system (Online optimization module, Test module, Manual optimization module), and they are shown as below:

a. Online optimization module

The main interface of online optimization part is as shown in Fig. 7.7, it consists of four parts of the data: (1) Data measuring part; (2) Data optimization part; (3) Optimization monitoring part; (4) Warning message part. Data measuring part displays the information of air, powder, oxygen and boiler efficiency, NO_x emission, etc., it aims to provide operators with current practical data of combustion status, air, powder, and oxygen is gathered from PI system, NO_x emission and CO concentration is gathered from online gas monitoring subsystem, boiler efficiency is calculated from the measured data. Data optimization part displays the parameter combination after optimization with SVM and GA, as well as the information of combustion status. The NO_x emission and CO concentration are calculated according to the optimization of operation parameters prediction, operation parameters like air and powder are set according to the optimization goal and optimization range. Optimization monitoring part shows the current setting of optimal conditions and goals, combustion status and data measuring, NO_x emission, etc., optimal conditions and goals can be set in "online setting" in the menu. Different optimal conditions and goals will lead to different optimization parameter combination. If the goals could not be satisfied, the module will give the best optimization value. Warning message part monitors the situation of air and powder, whether the wall temperature of superheater is over temperature, etc. Once the abnormal situation is found, an alarm message will be shown in the bar, and operator can deal with the abnormal situation for the first time. The function of loading, online database connection, running the reference parameters, and history data offering is set in the menu options, which will realize some auxiliary functions of the online optimization, including the database connection, parameters monitoring, optimization goal setting, parameter optimization, etc. (Fig. 7.8).

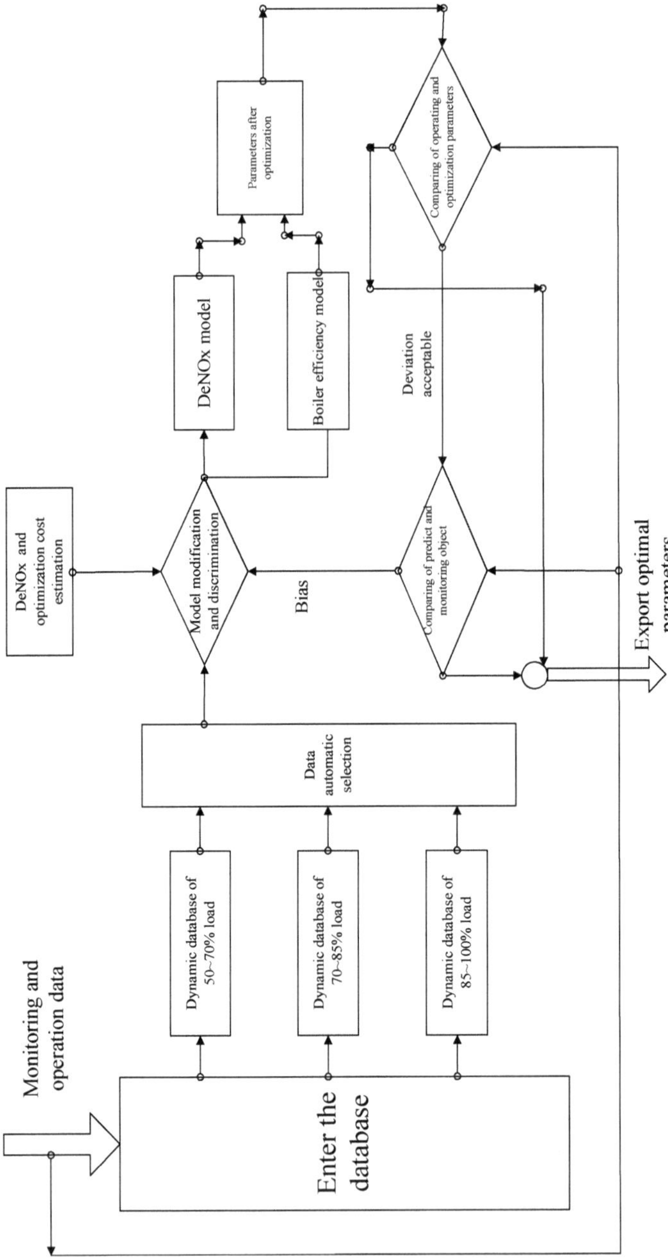

Fig. 7.5 Flowchart of online gas monitoring system and combustion optimization system

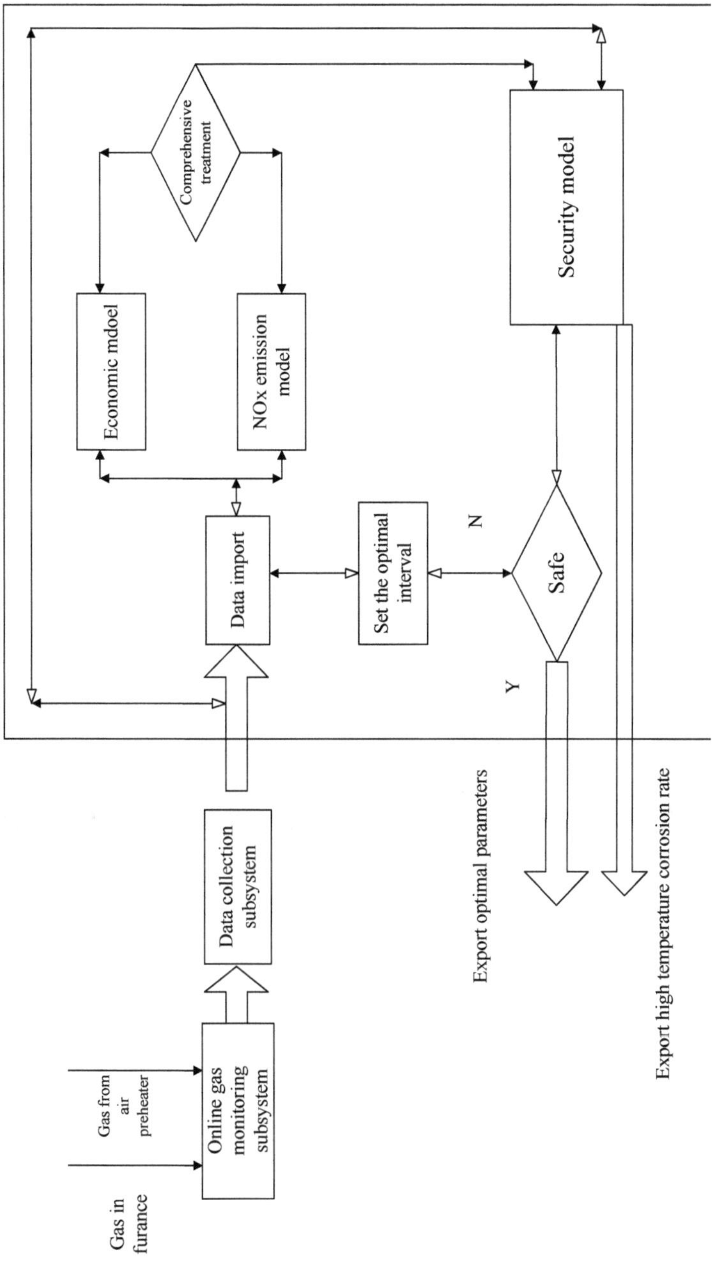

Fig. 7.6 The online optimization process

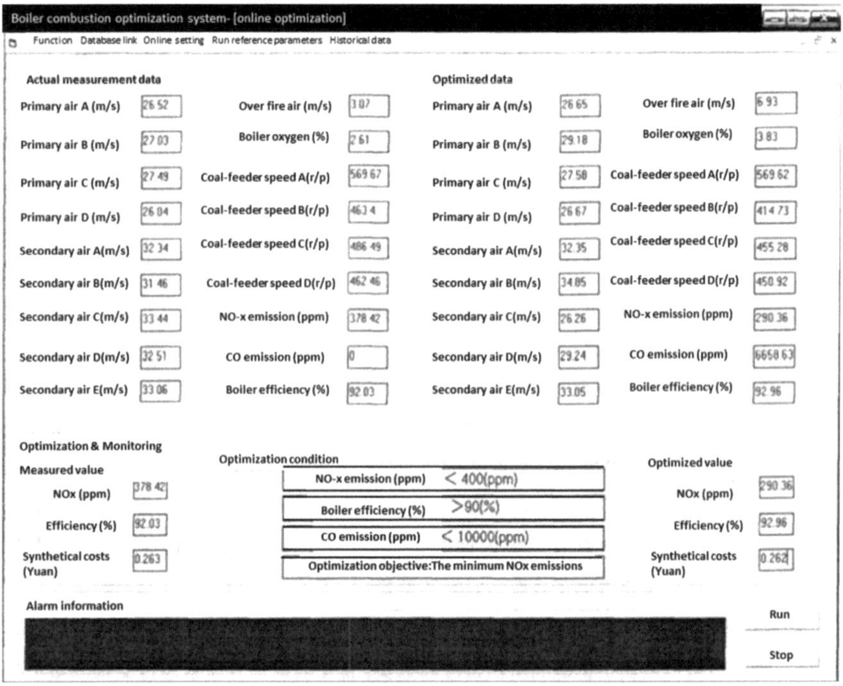

Fig. 7.7 The main interface of online optimization part

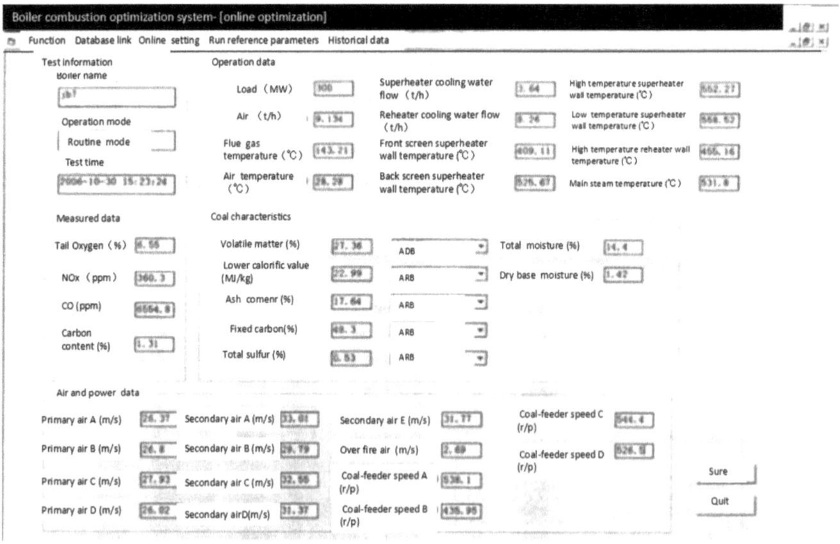

Fig. 7.8 The test data entry interface

b. Experiment module

The experiment part is mainly for power plant in or after adjustment test. The experiment consists of two parts: entry of test data and modeling of experimental data.

Entry of test data records provides manual data input interface for the condition that data could not be recorded automatically. Test data is stored in the database of the system in terms of independent test data for every test condition. This database is the source for modeling. The test data entry interface is shown in Fig. 7.6.

Experimental condition modeling interface, shown in Fig. 7.9, is designed mainly for boiler combustion adjustment experiment, which is conducted before the new unit is put into operation or after boiler renovation. The experimental model can examine whether optimization can be improved as well as lay the foundation for subsequent online combustion optimization. During the modeling process, operators can choose different modeling conditions to highlight the pertinence of model, and implement the optimization through the "manual optimization".

c. Manual optimization module

Manual optimization module is mainly aimed to solve the situation there is none in the experimental conditions or online database for optimization. Its data is divided into necessary optional parameters and the optional parameters. The necessary optional parameters are non-adjustable parameters, and combustion is implemented based on these non-adjustable parameters; optional parameters are the

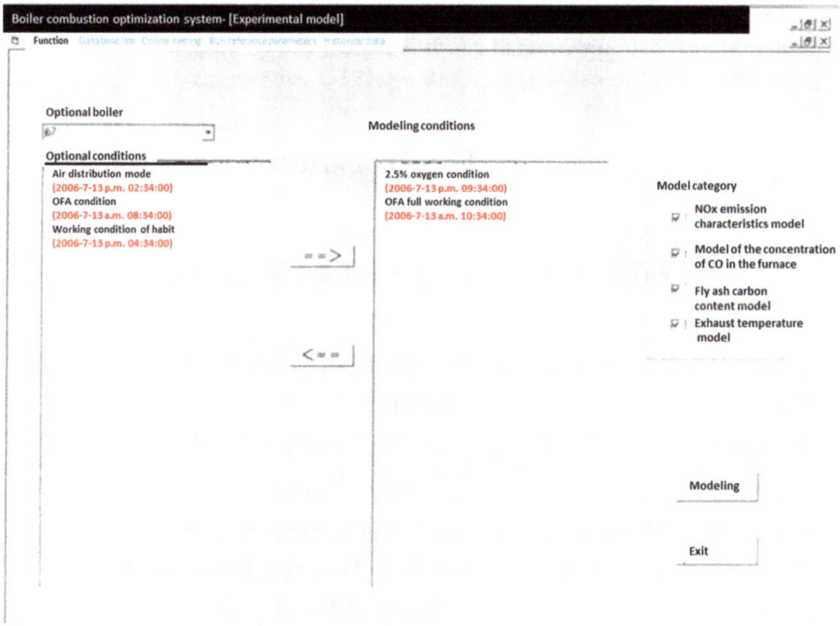

Fig. 7.9 The experimental condition modeling interface

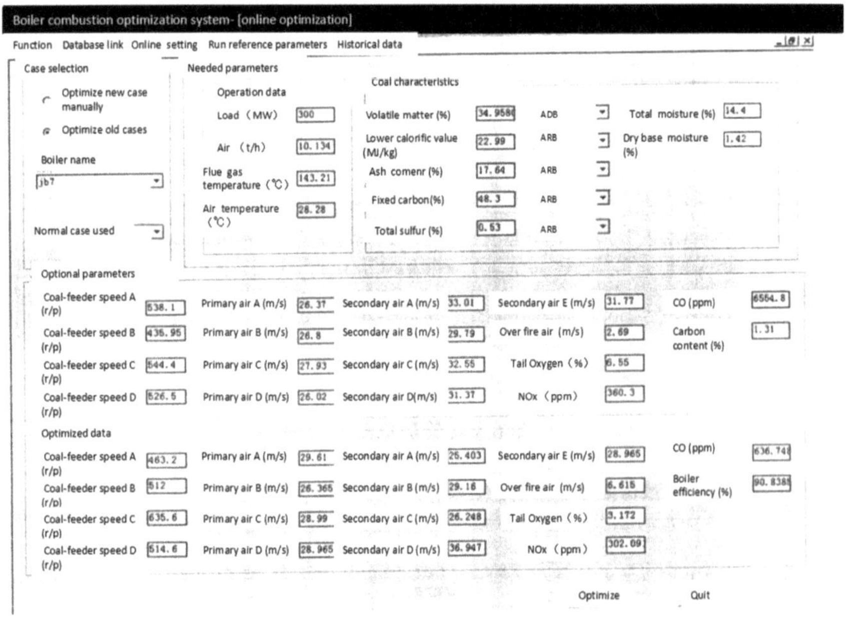

Fig. 7.10 Manual optimization module interface

ones through which combustion is conducted. When the range of model optimization parameters is determined, the original parameters can be discarded, but we keep the original data for comparison. It is shown in Fig. 7.10.

7.5 The Application of Online Combustion Optimization System

7.5.1 *Train and Prediction*

The data of operating conditions is collected from adjustment test with this software. We collected 670 cases totally before the formal test including NO_x emission, the fume temperature, and relevant parameters.

Before the formal experimental simulation, NO_x emission and fume temperature are built online with this software, and the 670 operating cases are predicted by the new models, prediction and the actual measured value contrast is shown in Figs. 7.11 and 7.12: It can be seen from the figure that, except for some points, the predicted value is in good agreement with the experimental value. The three maximum error in NO_x prediction is 36.36, 23.5, and 20.27%, while the others error stays small. The overall average error is 2.47%, absolute error variance is 63.44. The average absolute error is 8.36. With respect to the overall situation of NO_x

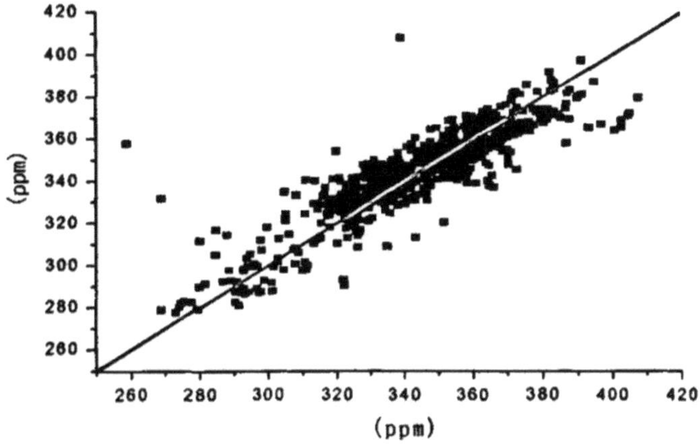

Fig. 7.11 SVM prediction and the actual measured value contrast of NO_x emission

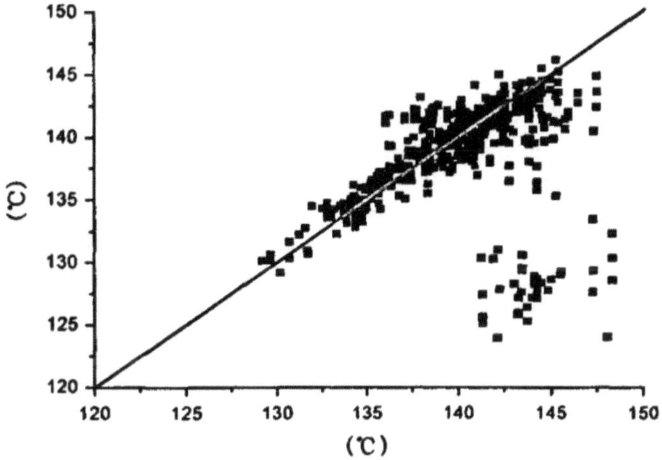

Fig. 7.12 SVM prediction and the actual measured value contrast of fume temperature

emissions, the average error of SVR prediction is 8 ppm, and 2.47% difference compared with the measured values. The NO_x emission predicting trends and values are accurate.

The fume temperature model is a little worse than NO_x model, there exit more than 10 cases in which the error is bigger than 10%, maximum error is 12.8% and average error is 1.57%, absolute error variance is 13.92, the average absolute error is 2.23. So the average error of SVR mode of fume temperature is about 2.23 °C, 1.57% difference compared with the measured values. The fume temperature predicting trends and values are accurate. These two models lay the foundation for the next step of formal acceptance tests and practical application.

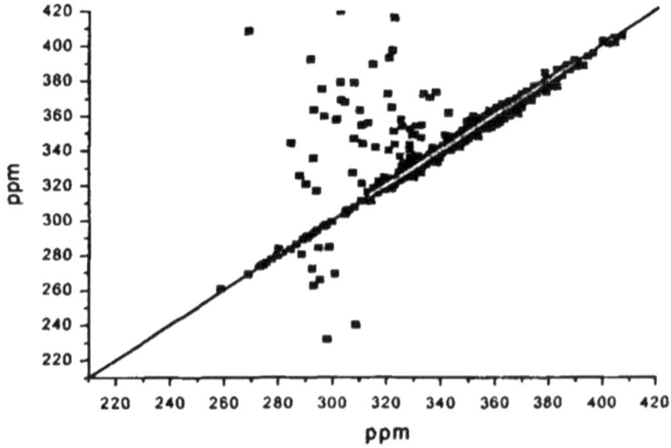

Fig. 7.13 BP prediction and the actual measured value contrast of NO_x emission

We also use BP neural network model to compare with SVR model in NO_x emission and fume temperature predicting. Their inputs and output are totally same. In BP neural network, 600 cases are selected as train data from 670 cases, train error is set at 5%. The BP prediction of all cases is shown in Figs. 7.13 and 7.14.

In NO_x prediction, the maximum error is 52% and average error is 1.77%, absolute error variance is 234.45, the average absolute error is 5.58. The BP neural network is worse than SVR from the perspective of error variance and maximum average error. Also, BP neural network is poor in generalization ability, the prediction error concentrates on the predict data set of 70 cases. If this predict dataset is removed from the prediction, the maximum error is 2.3% and average error is 0.47%, absolute error variance is 2.2, the average absolute error is 1.64. This means

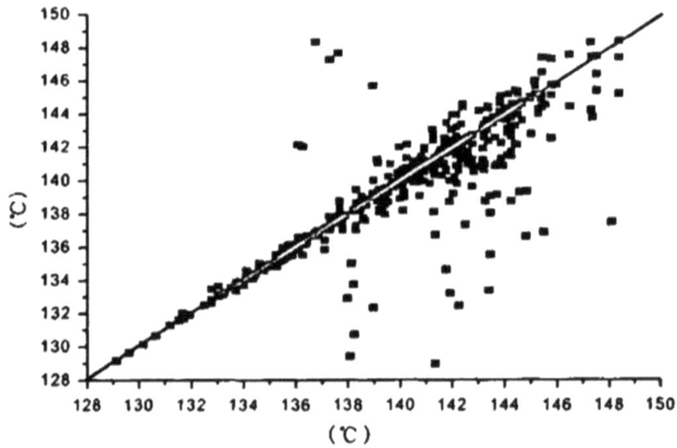

Fig. 7.14 BP prediction and the actual measured value contrast of fume temperature

the BP neural network prediction of train data is very good. However, it does badly in the predict dataset of 70 cases: the average error is 12.8%, absolute error variance is 978.83, which means the poor generalization ability of BP neural network. The comparison of BP neural network and SVR fume temperature model is similar. The generalized error in BP neural network prediction is due to our focus on reducing empirical risk and ignoring the generalization ability of the model.

From the prediction of the models above, SVR is superior to BP on the generalization ability. This is mainly because support vector machine algorithm takes minimizing structural risk as a training strategy, generalization ability is taken into account in the process of training, so the SVM model has a better prediction accuracy. This guarantees the successful application of SVM in many fields.

The power plant online combustion optimization software is applied and test on #7 boiler of a certain plant after developed. Unfortunately, online gas analysis system was not installed, we only acquired NO_x emission data, CO concentration cannot be collected. So CO concentration is not considered in optimization objects. The test process is as follows.

7.5.2 Test Purpose

By the online monitoring of boiler furnace flue gas composition and test of combustion optimization system, the purpose of the test is to master the situation of the system continued operation.

7.5.3 Test Condition

1. Boiler efficiency of basic conditions and NO_x emission test.
2. Low NO_x emission optimization test based on basic conditions, set parameters combination, and parameters of combustion state evaluation.
3. Combustion adjustment according to low NO_x emission optimization results and compared with the NO_x emission measured value after adjustment.
4. Conducted calculation of boiler efficiency in order to improve the boiler combustion optimization, set parameters combination, and parameters of combustion state evaluation.
5. Combustion adjustment according to high boiler efficiency optimization results and compared with the NO_x emission and boiler efficiency measured value after adjustment.

7.5.4 Test Data

The test data is shown in Table 7.1.

Table 7.1 Experiment data

Case	NO_x	Boiler efficiency	Load	Primary air speed				Second air speed					Wind speed	O_2
	(mg/m^3)	(%)	(WM)	I	II	III	IV	A	B	C	D	E	(m/s)	(%)
1	708.5	92.41	310	28.6	26.8	27.9	26.0	35.0	29.8	32.6	31.4	31.7	2.9	3.4
2	580.2	92.46	/	26.6	29.2	27.6	26.7	29.4	28.9	26.3	29.2	33.0	6.9	2.8
3	576.1	92.3	309	27.0	25.9	27.7	26.5	28.6	28.0	26.0	29.0	33.4	6.9	2.7
4	655.4	92.84	/	26.5	26.8	25.4	28.5	29.6	25.8	31.7	32.6	35.3	7.1	3.1
5	659.3	92.91	308	26.3	26.0	25.6	28.0	30.1	25.2	33.4	32.6	34.8	8.2	3.2

Where condition 1 is the basic one; condition 2 is the predicted one after de-NO_x optimization, condition 3 is the actual operation condition according to condition 2; condition 4 is the predicted one after boiler efficiency optimization, and condition 5 is the actual operation condition according to condition 4.

7.5.5 Result and Analysis

As shown in Table 7.1, the NO_x emission of basic operation condition is 708.5 mg/m^3, the boiler efficiency is 92.41%. After de-NO_x optimization, the predicted NO_x emission is 580.2 mg/m^3, the boiler efficiency is 92.46%. The actual NO_x emission after operation adjustment is 576.1 mg/m^3, and the boiler efficiency is 92.30%, the relative error of predicted and measured NO_x emission value is 0.71%, the relative error of predicted and actual boiler value is 0.11%. Therefore, it can be concluded that the NO_x emission and boiler efficiency SVM models agree with the physical aspect of the NO_x formation and boiler efficiency in combustion process; emission reduction effect is obvious, NO_x emission is derived to 18.8% reduction.

After boiler efficiency optimization, the predicted NO_x emission is 655.4 mg/m^3, the boiler efficiency is 92.84%. The actual NO_x emission after operation adjustment is 659.3 mg/m^3, and the boiler efficiency is 92.91%, the relative error of predicted and measured NO_x emission value is 0.08%, the relative error of predicted and actual boiler value is 0.6%. This means predict values show comparatively good agreement with the measured ones, and boiler efficiency increased by 0.5% compared to a basic condition. This result is accepted by the technicians of the power plant.

7.6 Summary

In this section, primitive principle and online application are introduced. We summarize the demand for the software system for the practical combustion optimization and the need for the local optimization.

Flame monitor system is analyzed, it is an important equipment for online optimization. Online algorithm is also introduced. Basic idea and process of online optimization software is discussed, and it has been applied in an industrial boiler. Tests show that the predicted values are comparatively in good agreement with the measured ones, and boiler efficiency increased by 0.5% compared to the basic condition.

Reference

1. Cauwenberghs G, Poggio T. Incremental and decremental support vector machine learning. Adv Neural Inf Process Sys, 2001: 409–415.

Chapter 8
Combustion Optimization Based on Computational Intelligence Applications: Future Prospect

Abstract In this section, the prospect of combustion optimization based on computational intelligence is forecasted. Combustion optimization is considered to be very significant for both the power plant de-NO_x and carbon burnout of fly ash. Then, three unclear factors are discussed and prospected: coal characteristics for model, dealing with large data set, and feature selection. Combustion optimization based on computational intelligence is a very valuable and practical technology to approve the industrial change.

Keywords Future prospect · Power plant de-NO_x · Carbon burnout
Coal characteristics for model · Dealing with large data set · Feature selection
New algorithm · Optimization model · Machine learning

Combustion optimization based on computational intelligence has been proposed for power plant de-NO_x and carbon burnout of fly ash application in this book, it is significant for pollution emission control and boiler efficiency. However, some factors are lack of reckon and boundedness of computational intelligence, perplexing combustion optimization. Future prospects are deserved to be taken into account.

(1) Coal characteristics for model. Although coal characteristics monitor system has been successfully put into use in power plant, real-time coal characteristics were not included in combustion optimization model. Coal quality like Aar, Var, Qar, net, and Mar, is used for model setup, as well as in optimal parameters sets searching. During the computation, coal quality is regarded as a constant, while it changes for two or three types sometime. But in power plant operation, coal characteristics transform unpredictably, leading to the invalidation of optimal parameters because of the computed model is based on a certain coal type. Online coal characteristics monitor signal delays, and boiler operating parameters in DCS do not correspond with coal monitor signal at the same time, this increases the difficulty in modeling the combustion with coal characteristic. Meanwhile, combustion in boiler is deferred than coal type change too. Fineness of pulverized coal is another important characteristic of

H. Zhou and K. Cen, *Combustion Optimization Based on Computational Intelligence*, Advanced Topics in Science and Technology in China,
https://doi.org/10.1007/978-981-10-7875-0_8

combustion, similar to coal monitor, measuring means are immature, leaving application in combustion optimization behind. How to add real-time coal characteristics and onlinc pulverized coal fineness into optimization model remain to be solved urgently.

(2) Dealing with large data set. Data set used in optimization model is restricted both in training data and testing data, its data patterns stay less than 1000 in most cases [1]. There has been a lot of interest on using kernels in various machine learning problems, with the SVM being the most prominent example in combustion optimization. SVMs are formulated as quadratic programming problems. The training time complexity of QP is $O(m^3)$ and its space complexity is at least $O(m^2)$. Dealing large operating data is time-consuming and it is not realizable on power plant application. Even the training data is collected from DCS, most of them are artificial selection in case of stumbling block in scaling data computation. Industrial data is fluctuant and uncertain in measuring period, artificial selection losses lots of combustion information and the selected data is subjective or the so-called average value, which impacts the result of human intervention. A lot of techniques have been studied to reduce the time and space complexity, popular and radical approaches are low-rank approximation, like greedy approximation, sampling decompositions; scale up kernel methods like chunking, sophisticated decomposition, others ways are Lagrangian SVM, reduced SVM, SMO, SVMlight, simpleSVM, etc [2, 3]. With this new model technique applied in dealing large combustion data, optimal conclusion will keep authenticity.

(3) Feature selection. Combustion is affected by many factors, such as coal character, boiler's load, air distribution, boiler style, burner style, furnace temperature, excess air ratio, pulverized coal fineness, the uniformity of the air and coal distribution, etc. These factors are analyzed logically by simplified mathematical formulas, human intervention in residuary element is implicit in these factors. Combustion is a very complex process influenced by massive factors, which is incarnated in DCS database, optimal models cover a certain number of elements into account, which is confined to the reasons influencing combustion we have known already. This selection process simplifies combustion model and results in unknown conclusion of optimization [4–6]. New algorithm taking enough factors into model is under researching.

References

1. Tsang IW, Kwok JT, Cheung PM. Core vector machines: Fast SVM training on very large data sets. J Mach Learn Res.; 2005:363–392.
2. Williams C, Seeger M. Using the Nyström method to speed up kernel machines. In: Advances in Neural Information Processing Systems. 2001. p. 13.
3. Platt J. Probabilistic outputs for support vector machines and comparisons to regularized likelihood methods. Adv Large Margin Classifiers. 1999;10(3):61–74.

4. Fung GM, Mangasarian OL. A feature selection Newton method for support vector machine classification. Comput Optim Appl. 2004;28(2):185–202.
5. Mao Q, Tsang IWH. Optimizing performance measures for feature selection. In: 2011 IEEE 11th International Conference on Data Mining (ICDM), IEEE; 2011. pp. 1170–1175.
6. Tan M, Wang L, Tsang IW. Learning sparse svm for feature selection on very high dimensional datasets. In: Proceedings of the 27th International Conference on Machine Learning (ICML-10). 2010. pp. 1047–1054.

Index

H. Zhou and K. Cen, *Combustion Optimization Based on Computational Intelligence*, Advanced Topics in Science and Technology in China,
https://doi.org/10.1007/978-981-10-7875-0

MIX
Papier aus verantwortungsvollen Quellen
Paper from responsible sources
FSC® C105338

If you have any concerns about our products,
you can contact us on
ProductSafety@springernature.com

In case Publisher is established outside the EU,
the EU authorized representative is:
Springer Nature Customer Service Center GmbH
Europaplatz 3, 69115 Heidelberg, Germany

Printed by Libri Plureos GmbH
in Hamburg, Germany